Ethical AI Development: A Global Case Study Approach

Azhar ul Haque Sario

Copyright

First Printing, 2025

Azhar.sario@hotmail.co.uk

ORCID: https://orcid.org/0009-0004-8629-830X
LinkedIn: https://www.linkedin.com/in/azharulhaquesario/

Disclaimer: This book is free from AI use. The cover was designed in Canva.

Copyright Disclaimer: The author has no affiliation with any government or regulatory board mentioned herein. This work is independently produced, and references to organizations, policies, and frameworks are made under the principle of nominative fair use for commentary and analysis.

Contents

Copyright ... 2

Introduction: Navigating the Global Landscape of AI Ethics 4

Part I: Foundational Frameworks and Governance Models 12

Germany - Data Privacy and Comprehensive Regulation (The GDPR Model) ... 12

China - AI-Powered State Surveillance and Social Credit .. 40

United States - Algorithmic Bias in Criminal Justice ... 50

Part II: AI in High-Stakes Societal Domains ... 62

United Kingdom - AI in Healthcare: Innovation vs. Patient Rights ... 62

Japan - AI and Robotics for an Aging Population ... 74

Estonia - AI and Democratic Processes: Combating Disinformation ... 86

Part III: Economic and Environmental Transformations ... 96

United Arab Emirates - AI and the Future of Work: Labor Displacement and Reskilling ... 96

South Korea - AI in Financial Services: Fairness in Lending and Credit ... 115

France - Generative AI and Intellectual Property in Creative Industries ... 135

Canada - AI and Environmental Sustainability: The Hidden Costs ... 149

Australia - Indigenous Data Sovereignty and AI ... 169

About Author ... 198

Introduction: Navigating the Global Landscape of AI Ethics

United States: Free Market and Innovation

In the United States, the approach to AI ethics is deeply rooted in its tradition of free-market capitalism and a strong emphasis on innovation. The prevailing philosophy is that less regulation often fosters more creativity and technological advancement. This doesn't mean ethics are ignored, but rather that the government prefers to let industry lead the way. Companies are encouraged to develop their own ethical guidelines and best practices. Think of it as a "let's see what works" approach. This has led to a vibrant and competitive AI landscape, with American tech giants pushing the boundaries of what's possible. However, this hands-off strategy also raises concerns. Critics worry that without stronger government oversight, issues like algorithmic bias and data privacy might not be adequately addressed, leaving consumers vulnerable. The debate continues to be a balancing act between fostering groundbreaking innovation and ensuring that this powerful technology is developed and used responsibly.

China: State-Driven Governance

China's approach to AI ethics presents a stark contrast to the United States. Here, the government takes a very active and central role. AI development is seen as a national priority, crucial for economic growth and global influence. The state sets the direction, invests heavily in research, and establishes clear rules for how AI should be used. This top-down approach allows for rapid, coordinated progress and the implementation of large-scale AI projects, such as smart cities and public surveillance systems. The emphasis is often on social harmony and national security. While this can lead to efficient and powerful applications of AI, it also brings up significant ethical questions, particularly around surveillance and individual freedoms. The government's ability to collect and analyze vast amounts of data creates a powerful tool for social management, sparking a global conversation about the balance between state control and personal privacy in the age of AI.

European Union: Rights-Centric Regulation

The European Union has positioned itself as a global leader in setting a high bar for AI ethics, with a strong focus on protecting fundamental human rights. The cornerstone of their approach is the idea that AI should be trustworthy and human-centric. This is not just a suggestion; it's being codified into law. The EU's AI Act is a landmark piece of legislation that categorizes AI systems based on their level of risk, with stricter rules for high-risk applications like those used in healthcare or law enforcement. The goal is to ensure that AI systems are safe, transparent, and accountable. This rights-based model prioritizes the individual, guaranteeing protections for data privacy, fairness, and human oversight. While some argue that this stringent regulatory environment could slow down innovation compared to the US or China, the EU believes that building trust is essential for the long-term, successful adoption of AI.

Canada: Public-Private Collaboration

Canada has carved out a unique path in the world of AI ethics by fostering deep collaboration between the public and private sectors. The government has invested significantly in creating AI research hubs and institutes, bringing together top academics, startups, and established companies. This collaborative ecosystem is built on the idea that the best way to tackle the complex ethical challenges of AI is by working together. The focus is on a "responsible AI" framework, which is developed through ongoing dialogue and partnership. This approach allows for a more nimble and adaptive response to the fast-paced world of AI. Instead of rigid, top-down regulations, Canada favors a model where ethical principles are co-developed and integrated into the entire AI lifecycle, from initial research to final deployment. This strategy aims to build an AI industry that is not only innovative but also deeply committed to ethical considerations from the ground up.

United Kingdom: Pro-Innovation Regulation

The United Kingdom is charting a course on AI ethics that it describes as "pro-innovation regulation." The goal is to create a regulatory environment that is flexible and adaptive, encouraging growth and experimentation without sacrificing safety and ethical standards. Rather than creating a single, overarching AI law, the UK is empowering existing regulators in different sectors—like finance, healthcare, and transportation—to develop their own context-specific rules for AI. This sector-by-sector approach is based on the belief that a one-size-fits-all solution isn't practical for such a diverse technology. The government sees its role as providing a high-level framework of principles, such as safety, transparency, and fairness, while letting the experts in each field figure out the details. This pragmatic approach aims to strike a balance, creating a system that is robust enough to build public trust but agile enough not to stifle the UK's burgeoning AI industry.

Japan: AI for a Harmonious Society

In Japan, the conversation around AI ethics is profoundly influenced by the concept of creating a harmonious and well-functioning society. The goal is not just to build powerful technology, but to integrate AI into daily life in a way that enhances social cohesion and supports an aging population. Japan's "Society 5.0" initiative envisions a future where AI and other advanced technologies work seamlessly with humans to solve societal challenges. This approach places a strong emphasis on building AI systems that are safe, reliable, and trustworthy. There is a deep cultural appreciation for technology that is helpful and respectful, acting as a partner to humanity rather than a disruptive force. The ethical framework in Japan seeks to ensure that AI contributes positively to the community, fostering a sense of shared well-being and social stability.

South Korea: Nurturing an AI Ecosystem

South Korea's strategy for AI ethics is closely tied to its ambition to become a global powerhouse in artificial intelligence. The focus is on nurturing a complete AI ecosystem, from fostering cutting-edge research and development to creating a supportive regulatory environment for businesses. The government is investing heavily in AI talent and infrastructure, viewing it as a key driver of future economic growth. Ethically, the approach is pragmatic, aiming to build public trust while ensuring that regulations don't hinder innovation. South Korea is actively working to establish clear guidelines for data usage, algorithm fairness, and accountability. The goal is to create a predictable and reliable framework that allows companies to innovate with confidence, knowing they are operating within ethically sound boundaries. This ecosystem approach aims to make South Korea not just a developer of AI, but a leader in responsible AI development.

Australia: A Risk-Based Framework

Australia is developing its approach to AI ethics around a practical, risk-based framework. The government has outlined a set of core principles for ethical AI, but it recognizes that not all AI systems carry the same level of risk. An AI used to recommend movies, for example, requires a different level of scrutiny than one used for medical diagnoses or criminal justice. Australia's approach, therefore, is to tailor the level of oversight to the potential for harm. This involves assessing the context in which an AI is used and applying stronger safeguards for high-stakes applications. The framework is designed to be a practical tool for businesses and organizations, helping them to identify, assess, and mitigate ethical risks throughout the AI lifecycle. This focus on risk management aims to foster responsible innovation by providing clear guidance on how to build AI that is safe, secure, and aligned with Australian values.

Singapore: A Practical, Industry-Focused Approach

Singapore has adopted a highly practical and industry-focused approach to AI ethics. As a global business hub, the nation's primary goal is to build a trusted and vibrant AI ecosystem that can drive economic growth. Singapore's "Model AI Governance Framework" is a prime example of this strategy. It's not a law, but rather a detailed guide for companies on how to implement AI responsibly. The framework provides practical steps and considerations for addressing issues like accountability, transparency, and fairness in a business context. The emphasis is on internal governance, encouraging companies to take ownership of the ethical implications of their AI systems. This business-friendly approach is designed to foster trust among consumers and international partners, positioning Singapore as a leading and responsible player in the global AI economy.

Brazil: Addressing Inequality

In Brazil, the dialogue on AI ethics is inextricably linked to the country's long-standing challenges with social and economic inequality. There is a strong focus on how AI could either help reduce these disparities or, if not carefully managed, make them even worse. Lawmakers and civil society groups are actively debating how to ensure that AI is developed and deployed in a way that is fair and inclusive. A key concern is algorithmic bias, particularly in areas like credit scoring or hiring, where biased systems could reinforce existing patterns of discrimination. Brazil is in the process of developing a national AI strategy and specific legislation that aims to protect citizens' rights and promote the use of AI for social development. The core of the ethical debate in Brazil is about harnessing the power of AI to create a more just and equitable society for all its citizens.

South Africa: Leapfrogging and Inclusion

For South Africa, the conversation around AI ethics is shaped by the concept of "leapfrogging"—using advanced technology to skip traditional stages of development and address persistent socio-economic challenges. The hope is that AI can help the country overcome issues related to healthcare, education, and service delivery. However, this optimism is tempered by a strong awareness of the digital divide and the potential for AI to exclude marginalized communities. Therefore, a central pillar of South Africa's approach to AI ethics is inclusion. The goal is to ensure that AI development is participatory and that the benefits of this technology are shared broadly across society. There is a focus on building local AI talent and creating solutions that are relevant to the African context, ensuring that AI serves as a tool for empowerment and not as a source of further inequality.

United Arab Emirates: A Vision for the Future

The United Arab Emirates has taken a bold and visionary approach to artificial intelligence, appointing the world's first Minister of State for AI. This move signals a deep commitment to making AI a central pillar of the nation's future development. The UAE's strategy is ambitious, aiming to use AI to transform government services, improve efficiency, and create a world-leading smart economy. The ethical framework is being developed in parallel with this technological push. The focus is on creating a positive vision for AI, where the technology is used to enhance human well-being and happiness. The government is actively working to establish clear guidelines on data privacy, transparency, and the responsible use of AI across all sectors. The UAE's approach is forward-looking, aiming to build a future where AI is not just a tool, but an integral part of a prosperous and well-governed society.

New Zealand: Indigenous Data Sovereignty

New Zealand brings a unique and vital perspective to the global conversation on AI ethics, with a strong focus on Indigenous data sovereignty. This concept is rooted in the rights of Māori, the Indigenous people of New Zealand, to have control over their own data. It asserts that Māori data—which can include everything from traditional knowledge and language to genetic information—should be managed in accordance with Māori values and principles. As AI systems are increasingly built on large datasets, the question of who owns and controls this data becomes critically important. New Zealand is pioneering a dialogue on how to build AI systems that respect and uphold Indigenous rights. This involves co-designing frameworks with Māori communities and ensuring that the use of their data is transparent, consensual, and beneficial to them. This focus on Indigenous data sovereignty is a crucial contribution to a more inclusive and equitable global approach to AI ethics.

Part I: Foundational Frameworks and Governance Models

Germany - Data Privacy and Comprehensive Regulation (The GDPR Model)

Introduction: Germany and the European Quest for Trustworthy AI

In the global conversation about how to manage the power of artificial intelligence, Europe has carved out a unique and profoundly influential path. It's a path that doesn't just focus on what AI can do, but on what it should do. At the heart of this movement is Germany, a nation that serves as both the economic engine of the European Union and a key architect of its regulatory philosophy. This chapter delves into Germany's role in championing the world's most comprehensive, rights-based approach to AI regulation. We will explore how the foundational principles of the now-famous General Data Protection Regulation (GDPR) have become the bedrock for the next frontier of digital governance: the landmark EU AI Act.

This isn't just a story about laws and directives. It's a story about values. The analysis here focuses on the intricate legal design and the very real, practical consequences of a framework built on a simple but radical idea: technology must serve humanity. This means placing the protection of fundamental human rights, the sanctity of personal data, and clear, unwavering accountability at the very core of AI development and deployment. This European model, often called the "Brussels Effect," is more than just a regional policy; it's setting a global gold standard for what it means to create trustworthy AI, compelling companies and countries far beyond its borders to take notice and adapt. But this journey isn't without its challenges. We will also investigate the deep-seated tensions that ripple through this model, particularly the delicate and often contentious balancing act between nurturing innovation and rigorously defending individual rights—a debate that is currently playing out in German policy circles with significant implications for the future of technology.

The GDPR as a Blueprint for AI Governance

Think of the General Data Protection Regulation (GDPR) as the constitutional foundation for Europe's digital world. When it arrived in 2018, it fundamentally changed the conversation about data. It wasn't just a set of rules; it was a declaration of digital rights. Principles that were once abstract legal concepts suddenly had teeth. Ideas like 'data minimization,' which means you should only collect the data you absolutely need for a specific purpose, became mandatory. The principle of 'purpose limitation' insisted that data collected for one reason, like shipping a package, couldn't be repurposed for another, like marketing, without clear consent. Most importantly, it empowered individuals with rights—the right to access their data, to correct it, and even to erase it. Germany, with its long-standing and culturally ingrained emphasis on privacy (think Datenschutz), was not just a participant but a leading voice in implementing and enforcing these rules.

Now, imagine extending that same protective logic to artificial intelligence. This is precisely what's happening. The principles of the GDPR are acting as the direct blueprint for AI governance. An AI system, especially a machine learning model, is incredibly hungry for data. It learns from the data it's fed. If the GDPR demands that data collection be fair and lawful, then any AI trained on that data inherits that legal obligation. You can't simply scrape the internet for photos to train a facial recognition model without considering the rights of every person in those photos.

The GDPR's emphasis on transparency becomes even more critical with AI. When a bank uses an AI algorithm to decide whether to grant you a loan, the GDPR's legacy, channeled through the new AI Act, demands an answer to the question: "Why?" You have a right to a meaningful explanation, to understand the logic behind the automated decision. This directly combats the "black box" problem, where even the creators of an AI can't fully explain its reasoning. In Germany, data protection authorities are already scrutinizing how companies use automated systems, setting precedents that ensure the rights established under the GDPR aren't rendered meaningless by opaque algorithms. The GDPR, therefore, wasn't just a data law; it was the essential groundwork for ensuring that the future of AI would be human-centric.

The EU AI Act: A Risk-Based Framework in the German Context

The EU AI Act is the next logical step, a sophisticated piece of legislation that builds directly on the GDPR's foundation. Instead of treating all AI as a single, monolithic thing, it introduces a brilliantly simple, yet effective, idea: a risk-based pyramid. It categorizes AI systems based on the potential harm they could cause to people's rights, safety, or well-being. This pragmatic approach allows for flexibility while being uncompromising on core values.

At the very top of the pyramid is Unacceptable Risk. These are AI systems considered a clear threat to people and are, quite simply, banned. This includes things like government-run social scoring systems that judge citizens based on their behavior, or AI that uses subliminal techniques to manipulate someone into doing something harmful. For a country like Germany, with its historical sensitivity to state surveillance and social control, this outright ban is a non-negotiable cornerstone of the regulation.

Below that is the largest and most scrutinized category: High-Risk AI. This is where the regulation really digs in. These aren't banned, but they are subject to strict requirements before they can ever reach the market. Think of AI used in critical infrastructure like the energy grid, medical devices that diagnose diseases, systems that recruit or promote employees, or algorithms used by judges to assist in sentencing. A German engineering company developing an AI for a self-driving car would fall squarely in this category. They would need to conduct rigorous risk assessments, ensure the data used to train the AI is high-quality and unbiased, maintain detailed logs of the system's performance, and provide clear information to the user. It's a heavy lift, but the goal is to ensure that when the stakes are high, the safeguards are even higher.

Further down the pyramid, we find Limited Risk AI. This category includes systems like chatbots. The main rule here is transparency. If you're talking to an AI, you should know you're talking to an AI. Similarly, if you're looking at a "deepfake" image or video, it must be clearly labeled as artificially generated. The goal is to prevent deception and empower users to make informed judgments.

Finally, at the base of the pyramid is Minimal Risk. This covers the vast majority of AI systems in use today, like the recommendation engine on a streaming service or the AI in a video game. The AI Act encourages these applications to voluntarily adopt codes of conduct but imposes no new legal obligations. This tiered structure is the Act's genius, allowing innovation to flourish in low-risk areas while putting formidable guardrails around applications that could truly impact human lives.

The 'Brussels Effect' and Germany's Global Influence

When the European Union, a market of over 450 million consumers, sets a high standard for a product, it rarely stays just a European standard. This phenomenon is known as the "Brussels Effect." Companies around the world are faced with a choice: either create two versions of their product—one for the highly regulated EU market and another for the rest of the world—or simply adopt the highest standard for everyone. More often than not, they choose the latter because it's simpler and more efficient. It also gives them a badge of trustworthiness they can market globally. With the GDPR, we saw American and Asian companies completely overhaul their privacy policies worldwide to comply. The exact same thing is happening with the AI Act.

Germany is a supercharger for this effect. As the world's fourth-largest economy and a global leader in industrial manufacturing, automotive engineering, and medical technology, what German companies do matters. When a corporate giant like Siemens, SAP, or a major car manufacturer like Volkswagen redesigns its AI systems to comply with the AI Act, it creates a ripple effect throughout its entire global supply chain. Suppliers in the United States, Japan, or India who want to continue doing business with these German titans must ensure their own AI components meet these stringent European requirements for transparency, data quality, and risk management.

This influence isn't just commercial; it's also ideological. By creating a clear, comprehensive legal framework for "trustworthy AI," Germany and the EU are providing a ready-made model for other countries to follow. Nations from Canada to Brazil to South Korea are looking closely at the AI Act as they draft their own regulations. They see a model that doesn't force a false choice between technological progress and democratic values. Instead, it argues that long-term, sustainable innovation is only possible when people trust the technology they are using. In a world searching for answers on how to govern AI, Germany, through the EU, is providing a very compelling and powerful one.

Navigating the Tension: Innovation vs. Rights in Germany

While the European model is lauded for its focus on human rights, it is not without its critics, and nowhere is this debate more alive than within Germany itself. The core tension is a classic one: how do you protect citizens without strangling the very innovation that promises to improve their lives and power the future economy? This is not just a theoretical debate; it's a practical struggle faced by German startups, researchers, and established industrial players.

On one side of the argument, you have Germany's strong civil society, data protection advocates, and trade unions. They point to the immense potential for AI to introduce bias into hiring, perpetuate discrimination in loan applications, and create opaque systems of control. For them, the strict rules of the AI Act are an essential and overdue defense of fundamental rights. They argue that true innovation is responsible innovation and that building trust from the ground up will ultimately be a competitive advantage, not a hindrance.

On the other side, you have powerful industry associations like the Bundesverband der Deutschen Industrie (BDI) and a vibrant startup scene, particularly in cities like Berlin and Munich. They express legitimate concerns that the high compliance costs and regulatory burdens of the High-Risk category could put them at a disadvantage against their less-regulated competitors in the US and China. A small German company developing a novel AI-powered diagnostic tool, for example, might face years of navigating complex approval processes and expensive audits, while a competitor elsewhere could potentially move faster to market. They worry that the regulation, while well-intentioned, could be a bureaucratic brake on the "Made in Germany" brand of technological leadership.

This balancing act is visible in national policy discussions. The German government has been a strong supporter of the AI Act's rights-based approach, but it has also pushed for "innovation-friendly" interpretations and the creation of "regulatory sandboxes." These are controlled environments where companies can test new AI technologies with real-world data under the supervision of regulators, allowing them to innovate without first having to clear every single regulatory hurdle. The outcome of this ongoing dialogue—between rigorously upholding rights and aggressively fostering innovation—will not only define the future of AI in Germany but will also serve as a crucial test case for whether the European model can deliver on both of its promises.

The EU's AI Rulebook: A Deeper Look at Its Moral Foundations

When we look at the European Union's approach to regulating artificial intelligence, it's easy to get lost in the technical details of the AI Act or the dense articles of the GDPR. But these documents are not just legal frameworks. They are the final chapters of a story that began long ago.

They are the modern expression of deeply held beliefs about people, power, and society. To truly understand why Europe is building its digital future this way, we need to look beyond the regulations and explore the powerful ideas that serve as their foundation. This philosophy is built on three core pillars: a belief in absolute duties, a cautious approach to unknown risks, and the unforgettable lessons of a painful history. Together, they explain why the EU sees AI not just as an economic opportunity, but as a profound challenge to its most cherished values.

Deontological Ethics: A Compass of Duties and Rules

At the very heart of the European regulatory mindset is a philosophical idea called deontology. In simple terms, deontological ethics argues that some actions are inherently right or wrong, regardless of their consequences. It's a philosophy based on duties and rules. Think of it like a universal moral code: you have a duty to tell the truth, not because it will always lead to the best outcome, but because lying is fundamentally wrong.

This way of thinking breathes life into the EU's major regulations. The General Data Protection Regulation (GDPR) isn't just a suggestion box for companies. It's a strict rulebook. It lays down clear duties for anyone who handles personal data. A company has a duty to collect only the data it absolutely needs for a specific purpose (data minimization). It has a duty to be transparent about what it's doing with that data. These aren't suggestions to be weighed against potential profits. They are moral obligations.

The EU AI Act follows the exact same logic. It doesn't just say, "Try not to cause harm with AI." Instead, it assigns specific, non-negotiable duties to the creators and users of AI systems, especially those deemed "high-risk." For example, a company developing an AI to help diagnose medical conditions has a strict duty to use high-quality, unbiased data. They have a duty to keep meticulous records. They have a duty to ensure a human can intervene at any time. The morality isn't judged by the success rate of the AI; it's judged by whether these fundamental duties were fulfilled. This rule-based approach creates a predictable and stable environment. It's a declaration that in Europe's digital world, there are clear lines that cannot be crossed, ensuring that fundamental principles are never sacrificed for a convenient outcome.

The Precautionary Principle: A Safety Net for an Unknown Future

The second major pillar of the EU's philosophy is the Precautionary Principle. This is a concept born from European environmental and public health law, and its core message is simple: better safe than sorry. It mandates that when an activity raises potential threats of harm to human health or the environment, protective measures should be taken even if some cause-and-effect relationships are not fully established scientifically. You don't wait for the disaster to happen to prove the danger existed. You build the guardrails on the cliff edge before the first car goes over.

This principle is the master architect of the AI Act's famous risk-based pyramid. The EU looked at the landscape of artificial intelligence and, using the precautionary principle, decided to act proactively against potential harms before they could become widespread societal problems.

At the very top of the pyramid are "unacceptable risk" systems. These are banned outright. AI systems that create subliminal manipulations or government-run "social scoring" systems fall into this category. The EU isn't waiting to see evidence of mass societal damage; the potential for harm to human dignity and freedom is so great that the principle demands they be stopped before they are ever deployed.

Just below that is the "high-risk" category. This is where the principle does its most important work. AI used in critical areas like medical devices, self-driving cars, judicial systems, or hiring processes must undergo strict conformity assessments before they can enter the market. This is precaution in action. It forces developers to prove their systems are safe, fair, and robust ahead of time, rather than asking for forgiveness after a faulty algorithm has denied thousands of people loans or produced biased parole decisions. This proactive, preventative mindset stands in stark contrast to more reactive models, reflecting a deep-seated belief that the first duty of governance is to protect its citizens from foreseeable harm.

History's Echo: Fundamental Rights as the Unshakable Bedrock

You cannot understand Europe's passion for fundamental rights without understanding its 20th-century history. The continent was the stage for totalitarian regimes that used state power and surveillance to crush the human spirit. The post-war European project was, in many ways, a solemn vow to say "never again." It was a commitment to building a political order where the dignity and rights of the individual would be the supreme purpose of the state, not the other way around.

This historical lesson is the direct source of the EU's unyielding focus on human rights in its technology policy. Data protection, for instance, is not seen as a mere consumer preference in Europe; it is enshrined as a fundamental right, on par with freedom of expression. The GDPR is therefore not just a data law; it is a human rights instrument designed to correct the power imbalance between the individual and large organizations. It's a digital shield against the kind of state and corporate overreach that history has shown can lead to terrible outcomes.

This commitment shines brightest in the AI Act's prohibitions. The ban on social scoring is a perfect example. The reason is not purely technical; it is deeply philosophical. A social score reduces a complex human being to a single, state-controlled number, stripping them of their intrinsic dignity. It is a tool of social control that is terrifyingly reminiscent of the very totalitarianism the European Union was built to oppose. Likewise, the strict rules for AI in law enforcement and the justice system are a direct reflection of historical lessons learned about the abuse of state power. For the EU, AI governance is not a neutral, technical exercise. It is a moral imperative to ensure that this powerful new technology is used to empower people, not to control them.

The German Debate: A Nation's Soul-Searching on AI

To see this philosophical tension play out in real-time, we only need to look at the ongoing debate within Germany, the EU's economic powerhouse. The country is currently grappling with a seemingly bureaucratic question: which government agency should be in charge of overseeing the AI Act? But this debate is not about org charts and office space. It is a proxy war for the soul of AI governance itself.

On one side, you have powerful voices, often from the economic and transport ministries, who argue that the

primary goal should be innovation and competitiveness. They advocate for a new, specialized agency that is business-friendly and can move quickly. Their deep-seated fear is that if Germany and Europe get bogged down in rights-based bureaucracy, they will lose the global AI race to the United States and China, jeopardizing future prosperity. They see AI primarily through the lens of economic opportunity and believe that regulation, while necessary, must not stifle growth.

On the other side are the data protection authorities and consumer rights groups. They argue that oversight should be handled by an existing body, like the Federal Data Protection Commissioner, which is already steeped in a culture of safeguarding fundamental rights. Their greatest fear is that a new, pro-innovation agency would prioritize economic gains over the protection of citizens. For them, AI is first and foremost a societal challenge that threatens to erode the rights-based legal order built so carefully over the past 70 years.

This German conversation is a perfect microcosm of the larger European challenge. It is the wrestling match between two valid but competing priorities: fostering economic strength and safeguarding foundational human rights. The final decision Germany makes will send a powerful signal. It will reveal whether, when faced with the immense pressures of technological competition, the continent will hold fast to its historical commitment to a rights-first approach or shift toward a model that gives more weight to economic pragmatism.

The GDPR's Enduring Blueprint for AI Governance

Imagine building a skyscraper. You wouldn't start with the gleaming windows or the rooftop garden; you would begin by digging deep and pouring a solid, unshakable foundation. In the world of European technology regulation, the General Data Protection Regulation (GDPR) is that foundation. The new, towering structure being built upon it is the EU AI Act. This new law, designed to govern the complex world of artificial intelligence, doesn't exist in a separate universe. Instead, it carefully and deliberately interlocks with the principles that have already reshaped how organizations handle personal data.

The GDPR gave us a powerful vocabulary for digital rights. It introduced core ideas like accountability, fairness, transparency, data minimization, and purpose limitation into the mainstream. These weren't just abstract legal concepts; they became practical requirements for businesses worldwide. The AI Act takes this established language and applies it to the next frontier of technology. It recognizes that at the heart of most powerful AI systems lies a vast ocean of data, much of which is personal. Therefore, governing the AI means strengthening the rules that govern the data it consumes.

This creates a wonderfully coherent, layered regulatory landscape. For any organization already putting in the hard work to comply with the GDPR, the journey towards AI Act compliance is not a leap into the unknown. It's the next logical step on a path they are already walking. The principles are familiar, the internal processes for data governance are already in place, and the mindset of putting data protection first provides a massive head start. For example, the GDPR's absolute insistence that you must have a legitimate, legal reason to process someone's personal data is the starting pistol for any AI development. Before a single line of code for a new AI model is written, the developers must first answer a very GDPR-centric question: do we even have the right to use this data in the first place?

Weaving GDPR Principles into the Fabric of the AI Act

The DNA of the GDPR is visibly present throughout the AI Act, with its core principles being not just referenced, but actively expanded and given new life in the context of automated systems. It's a direct evolution of thought, moving from protecting data in databases to protecting people from data-driven decisions.

Let's start with fairness and transparency. Under the GDPR, this means you must be open about what data you are collecting and how you are using it, and you must not use it in a way that is unjustifiably detrimental to the individual. The AI Act takes this principle and turbocharges it for the age of algorithms. For a high-risk AI system, like one deciding who gets a loan, transparency isn't just about a privacy policy. It's about explainability. The provider must be able to shed light on how the AI reached its decision. This prevents "black box" scenarios where even the creators can't explain the logic, ensuring that decisions are not just efficient, but also fair and scrutable.

Then there's data minimization and purpose limitation. The GDPR taught organizations a crucial lesson: stop hoarding data. You should only collect the data you absolutely need (minimization) for the specific, declared reason you collected it for (purpose limitation). This philosophy is a direct counter to the old AI development mantra of "more data is always better." The AI Act refines this by focusing on data quality. It specifies that for high-risk systems, the data used to train, validate, and test the AI must be relevant, representative, and as free of errors and biases as possible. It's a shift from a mindset of data quantity to one of data quality and integrity, ensuring that an AI trained to serve a diverse population isn't fed data from only one demographic, which would inevitably lead to biased and unfair outcomes.

Accountability Amplified: From Records to Comprehensive Blueprints

The principle of accountability is arguably where the link between the two regulations is strongest and most consequential. The GDPR made accountability a central pillar, demanding that organizations not only comply with the law but also be able to demonstrate their compliance. The primary tool for this is Article 30, which requires maintaining detailed Records of Processing Activities (ROPAs). A ROPA is like a meticulous logbook, detailing what data you process, why you process it, and who you share it with.

The AI Act takes this concept of a logbook and transforms it into a full set of engineering blueprints. It extends the obligation far beyond just recording data activities. For providers of high-risk AI systems, it mandates a suite of documentation that is vastly more detailed. Article 17 demands the implementation of a comprehensive Quality Management System, a formal system that documents processes and responsibilities for achieving quality policies and objectives throughout the AI's lifecycle. Think of this as the overall governance framework and rulebook for the AI.

Then, Article 18 requires extensive technical documentation. This isn't a high-level summary; it's a deep dive into the very soul of the AI system. It must explain the system's capabilities and limitations, the logic behind its design, and the technical methods used to build it. Crucially, it must detail the datasets used for training, including their origin, scope, and main characteristics. This documentation must be so thorough that it allows authorities to assess the system's compliance. In essence, while the GDPR asked, "What are you doing with the data?", the AI Act asks, "Show us exactly how you built the machine that uses the data, prove to us that it's safe, and explain its every decision." This is a monumental leap in accountability.

A Cumulative Approach in Action: The Real-World Test

To see how this layered regulation works in practice, let's revisit the example of a German company developing a high-risk AI system for screening job applicants. This company isn't starting from scratch; it has been living in a GDPR world for years.

Phase One: The GDPR Foundation. Because they process applicant data, they already have a solid GDPR compliance program. They have appointed a Data Protection Officer (DPO) who oversees their data strategy. They maintain a detailed ROPA for their HR processes. Before ever considering an AI, they conducted a Data Protection Impact Assessment (DPIA) to identify and mitigate risks associated with processing sensitive applicant information. They have a clear, lawful basis for using this data and established procedures for handling data subject rights, like the right to access or delete one's data. All of this work is essential, mandatory, and already done.

Phase Two: Building with the AI Act. Now, as they introduce a high-risk AI to streamline screening, the AI Act's requirements click into place on top of their existing GDPR framework. Their DPO's role expands to oversee AI compliance. The DPIA they conducted becomes a foundational document for the new, more intensive risk assessments required by the AI Act. But new tasks emerge. They must now create the extensive technical documentation required by Article 18. This means detailing the AI model's architecture, its underlying logic, and the specific measures taken to test for and mitigate bias. They can't just use their existing applicant data; they must rigorously analyze it to ensure it is representative, complete, and doesn't contain historical biases that the AI would learn and perpetuate. They must set up the Quality Management System under Article 17, which includes post-market monitoring to ensure the AI continues to perform fairly and accurately once deployed.

This demonstrates the cohesive, cumulative power of the EU's approach. The effort invested in GDPR compliance is not redundant; it is the prerequisite for the next stage of responsible innovation. It ensures that the conversation around AI doesn't begin with what's technologically possible, but with what is fundamentally fair, transparent, and accountable.

A Risk-Based Approach to Regulation: A Common-Sense Framework

Imagine trying to apply the same set of safety rules to both a child's tricycle and a high-speed racing motorcycle. It wouldn't make any sense. The level of danger is completely different, so the regulations should be too. This is the simple, powerful idea at the heart of the European Union's AI Act, a framework Germany has embraced as it navigates the complex world of artificial intelligence. Instead of a clumsy, one-size-fits-all approach, the EU has created a tiered system that smartly tailors the rules to the potential risk an AI system poses to our health, our safety, and our fundamental rights.

This risk-based ladder has four main rungs, each with its own set of expectations.

At the very top, you find the Unacceptable Risk category. These are the AI applications that are considered so dangerous to our society and individual freedoms that they are banned outright. Think of the kinds of technologies you might see in a dystopian film. This includes government-led social scoring systems that rank citizens based on their behavior, or AI designed to subtly manipulate people into making choices they wouldn't otherwise make, effectively hijacking their free will. It also includes most uses of real-time biometric scanning (like facial recognition) in public places by police, with only very narrow and specific exceptions for grave threats. The message here is clear: some technological doors are best left unopened.

The next rung down is the High-Risk category. This is where the most intense regulatory scrutiny happens. These aren't banned systems, but they are powerful ones that operate in critical areas of our lives. We're talking about AI used in essential infrastructure like our power grids, AI that helps doctors make medical diagnoses, AI that sorts through job applications, or AI used by judges and police in the justice system. Because a mistake in these areas could have severe consequences, the AI Act demands a lot from their developers. Before a high-risk AI can be sold in the EU, it must go through a tough checklist. This involves rigorous risk assessments, ensuring the data used to train it is high-quality and unbiased, creating detailed technical documentation, enabling the system to log its operations, and most importantly, guaranteeing meaningful human oversight. These systems must be built to be robust, secure, and transparent.

Stepping down again, we arrive at the Limited Risk category. The key word here is transparency. These AI systems don't pose a direct threat to our safety, but they do have the potential to deceive us. The classic examples are chatbots or the increasingly common "deepfakes." The law doesn't seek to ban these tools, but it does insist on a simple rule: you, the user, must be clearly informed that you are interacting with an AI, not a person, or viewing AI-generated content. This allows you to make an informed decision about how to proceed. It's about maintaining honesty and trust in our digital interactions.

Finally, at the bottom of the ladder, is the Minimal or No Risk category. This is where the overwhelming majority of AI applications live. Think of the AI that recommends movies on a streaming service, the spam filter in your email, or the smart opponents in a video game. The AI Act essentially leaves these applications alone, allowing for their free use and development without burdensome regulation. The idea is to let innovation flourish where the risks are negligible, though companies are encouraged to voluntarily adopt codes of conduct as a best practice.

This layered approach is brilliantly practical. For instance, in a German hospital, an AI tool that analyzes medical scans to spot early signs of cancer would absolutely be a high-risk device. It would need to be thoroughly tested and certified, perhaps by an independent body, before a single patient's scan could be run through it. Meanwhile, the simple chatbot on that same hospital's website, used just for booking appointments, would fall into the limited-risk bucket. Its only real obligation would be to introduce itself as an AI. This system ensures that regulation is a targeted shield, not a suffocating blanket.

The Tension Between Innovation and Rights Protection: A German Balancing Act

Within Germany, a country renowned for both its engineering prowess and its deep-seated commitment to individual privacy, the rollout of this comprehensive AI regulation has sparked a profound and necessary debate. It's a classic tug-of-war, a struggle to find the perfect balance between two crucial goals: fostering a vibrant, competitive technology sector and building an unshakeable fortress around the fundamental rights of its citizens. The core question is this: are strict rules a barrier that stifles innovation, or are they the very foundation upon which a trustworthy—and ultimately more successful—AI ecosystem can be built?

On one side of this debate are the proponents of a more lenient, innovation-first approach. They look at the speed and scale of AI development in the United States and China and worry that Europe is at risk of being left behind. From their perspective, every regulatory hurdle, every compliance checklist, is a potential delay that could put German and other European companies at a serious competitive disadvantage. They argue that overly strict interpretations of the rules could bury promising startups in paperwork, scaring away investment and pushing brilliant minds to move elsewhere. Their goal is to unleash the power of AI to drive economic growth and solve major challenges.

On the other side stand Germany's formidable data protection authorities (DPAs) and civil liberties organizations. Their starting point is that fundamental rights, such as privacy and non-discrimination, are non-negotiable. They contend that any rush toward innovation that compromises these principles is a false economy. In fact, they argue the opposite: that the EU's focus on creating "trustworthy AI" will become a major competitive advantage in the long run. In a world increasingly wary of technological overreach, a "made in Europe" label that guarantees safety, fairness, and ethical design could become a globally recognized seal of quality that consumers and businesses actively seek out.

This ideological conflict was thrown into sharp relief by a proposed law that emerged from Germany's Federal Ministry for Digital and Transport in 2025. The proposal was a bold move. It sought to change who supervises high-risk AI systems in very sensitive fields like law enforcement and border security. Instead of the state data protection authorities, who have long been the guardians of citizen rights, the ministry wanted to give this oversight role to the Federal Network Agency. The ministry's stated reason was to reduce what it saw as "barriers to innovation," hinting that the DPAs were too focused on rights and not enough on progress.

This suggestion was met with a swift and unified backlash from the DPAs. They issued a joint statement arguing that the plan was not just a bad idea, but a dangerous one. They claimed it would significantly weaken the protection of fundamental rights, as the Federal Network Agency's expertise lies more in technical and economic regulation than in constitutional law. Furthermore, they argued it went against the very spirit of the EU AI Act, which envisions a central role for data protection experts.

The debate over an AI system for predictive policing perfectly captures this struggle. Who is better equipped to judge whether such a tool is fair and constitutional? The data protection authorities, with their decades of experience in assessing impacts on fundamental rights? Or a technical agency, potentially more aligned with the government's goals of efficiency and technological deployment? This is more than a bureaucratic turf war; it's a fight to define the very purpose of AI governance. Is the primary goal to accelerate innovation at all costs, or is it to carefully steer that innovation within the firm guardrails of a rights-based democracy?

Extraterritoriality and the "Brussels Effect": How EU Rules Go Global

The influence of the EU's AI Act doesn't stop at the borders of Germany or the European Union. Much like its famous predecessor, the General Data Protection Regulation (GDPR), the AI Act has a remarkable global reach built into its very design. This principle is called extraterritoriality, and it works in a straightforward way: the rules don't just apply to companies based in the EU. They apply to any company anywhere in the world that wants to sell its AI system on the EU market or whose AI's output is used within the EU. This simple provision triggers a powerful phenomenon known as the "Brussels Effect."

The Brussels Effect describes the way EU laws and standards often become the default global standards. The European Union is one of the largest and wealthiest single markets on the planet. For most international technology companies, choosing not to do business there is simply not an option. Faced with the choice of creating a special, highly-regulated version of their product just for Europe and a different, less-regulated version for everyone else, most companies make a simple business calculation. It is far more efficient, predictable, and cost-effective to just build one product that meets the highest standard—which is often the EU's—and then sell that single version globally.

In this way, the EU doesn't need to impose its rules on other countries. The magnetic pull of its market does the work for it. This dynamic effectively exports European values on technology and ethics around the world. The AI Act's strict requirements for transparency, risk management, data quality, and human oversight are no longer just a European concern. They become a practical checklist for developers in North America, Asia, and beyond who want to access European customers.

This forces a worldwide convergence toward the EU's rights-centric and human-first model of AI development. It becomes a potent tool of regulatory influence, shaping the international tech landscape without a single treaty being signed. It subtly nudges the entire global industry toward a higher standard of care and responsibility.

Consider the example of a promising startup in Canada that has developed a sophisticated, AI-driven algorithm for credit scoring. The company sees a huge opportunity to sell its system to major banks in Germany and France. To do so, they now have to ensure their product can pass the EU AI Act's rigorous conformity assessment for high-risk systems. Their engineers and data scientists must document their data sources, prove their algorithm is not discriminatory, build in mechanisms for human review, and meet high standards for cybersecurity.

Having invested all that effort to make their product EU-compliant, what is the Canadian startup likely to do when it looks to sell to markets in the United States or Australia? It is highly unlikely they will build a second, lower-standard version. Instead, they will market their system as meeting the world's toughest regulations, turning a compliance necessity into a selling point. The high standards of the EU AI Act, adopted to protect a German bank customer, have now been embedded into a product used in Canada and beyond. This is the Brussels Effect in action, a quiet but powerful force pulling the global trajectory of technology toward a more responsible future.

The Art of the AI Diary: Comprehensive Documentation

Imagine you're building the most intricate, beautiful clock the world has ever seen. You wouldn't just assemble it from a pile of gears and springs and hope for the best. You would have detailed blueprints, notes on where each tiny component came from, a log of every adjustment you made, and a record of how accurately it keeps time. This is precisely what comprehensive documentation is for an AI system in the European market. It's not just tedious paperwork; it's the system's biography.

This biography must tell a complete story. It starts with the data sources. Where did you get the information to train your AI? Was it gathered with consent? Is it representative, or does it have hidden biases? Regulators want to see the birth certificate of your data. Think of it as the AI's family tree, proving its lineage is legitimate and sound.

Next comes the model's architecture—the actual blueprint. How does the AI think? What logic does it use? You don't need to reveal every single secret, but you must be able to explain its fundamental design. This is followed by the training procedures, which are like the AI's school years. What lessons was it taught? What textbooks (training data) did it study? How did you test it along the way to make sure it was learning the right things?

Finally, you need the testing results—the AI's final exam scores. How did it perform in different scenarios? Where did it excel, and more importantly, where did it fail? For systems deemed "high-risk," like those used in hiring or medical diagnostics, this documentation isn't a "nice-to-have." It is the central pillar of trust. If your AI makes a critical decision, you need to be able to open this diary and trace the steps, explaining exactly how and why it reached its conclusion. It's your ultimate defense and your proof of due diligence.

Fortune Telling for Prudence: The Role of Impact Assessments

Before a major construction project begins, engineers conduct exhaustive studies. They analyze the soil, model for earthquake resistance, and assess the environmental impact. They don't wait for the skyscraper to start leaning to wonder if they should have checked the foundation. A Fundamental Rights Impact Assessment for a high-risk AI system is the digital equivalent of this critical foresight. It's a mandatory exercise in structured imagination, asking one simple, powerful question: "What could go wrong?"

This isn't about vague fears. It's a systematic process of identifying potential harms before your AI system ever interacts with a real person. You are forced to put on different hats. As a privacy advocate, you might ask, "Could this system inadvertently reveal sensitive personal information?" As a social justice expert, you'd question, "Could this AI, trained on historical data, perpetuate biases and discriminate against certain groups?" For example, an AI designed to approve loans could unfairly penalize applicants from specific neighborhoods if its training data reflects past discriminatory lending practices.

The goal is to map out these risks. Once a potential harm is identified, you have to mitigate it. This means building in safeguards, tweaking the algorithm, or perhaps even deciding that a particular use of the AI is simply too dangerous to deploy. It's a proactive search for problems. This process transforms ethics from a philosophical debate into a practical engineering discipline. It ensures that the profound respect for individual rights, so central to German and EU law, is not just a policy on a shelf but a tangible, structural component of the technology itself. It's about building a system that is safe by design, not by accident.

Curating the Source: The Necessity of Robust Data Governance

An AI model is, at its core, a reflection of the data it was trained on. You can have the most brilliant algorithm in the world, but if you feed it a diet of junk data, you will get junk results. This is the timeless principle of "garbage in, garbage out." That's why robust data governance isn't just a compliance task; it is the single most important factor determining the quality and fairness of your AI. Think of yourself as the curator of a world-class museum. You wouldn't just accept any old painting; you would meticulously verify its authenticity, condition, and origin.

This curation process begins with data collection. Under the GDPR, this means data must be collected lawfully, fairly, and transparently. People need to know why their data is being gathered and consent to its use. This principle forms the ethical bedrock of your entire system.

Next is labeling. When you're training an AI, you're essentially showing it thousands of examples and telling it, "This is a cat," "This is a dog." If your labels are wrong or reflect subtle biases, your AI will learn those same mistakes. For instance, if data used to train a hiring AI consistently labels male candidates as "strong leaders," the AI will learn to favor men, even if that bias is unintentional. Establishing clear frameworks for accurate and fair labeling is therefore crucial.

Finally, there's data management. This involves keeping the data secure, ensuring its integrity, and managing its lifecycle. The AI Act places enormous emphasis on data quality because regulators understand that this is where many potential harms, like bias and inaccuracy, originate. A strong data governance framework is your quality control, ensuring the foundational ingredient of your AI is pure, compliant, and fit for purpose.

The Round Table: Why Cross-Functional Collaboration is Non-Negotiable

In the past, building a new piece of technology often followed a linear path. The engineers and developers would build something brilliant in their isolated workshop. Then, they would pass it over to the legal and compliance departments, almost as an afterthought, to get it "approved" before launch. That model is now completely obsolete, especially in the tightly regulated European AI landscape. Today, building a compliant and ethical AI system is more like filming a blockbuster movie. You need everyone—the director, the writer, the effects artists, and the legal advisors—in the room from day one.

AI development can no longer be a siloed activity. The technical teams, who understand the "how," must be in constant conversation with the legal experts, who understand the "what is permissible." The compliance officers, who know the specific articles of the GDPR and the AI Act inside and out, need to work alongside the developers, not audit them after the fact.

Perhaps most importantly, ethics experts must have a seat at the table from the very beginning. They are there to ask the "should we?" questions that go beyond mere legal compliance. Just because you can build something, does that mean you should? What are the potential societal impacts? This continuous dialogue creates a virtuous cycle. A lawyer might point out a privacy risk in the data collection plan, prompting the technical team to find a more privacy-preserving alternative. An ethicist might question the fairness of an algorithm's outcome, leading the team to seek out more diverse training data.

This collaborative approach isn't about slowing down innovation. It's about making it smarter, safer, and more resilient. It prevents costly mistakes and ensures that the final product is not only technologically impressive but also socially responsible and legally sound from its very core.

China - AI-Powered State Surveillance and Social Credit

China's Digital Gaze: Unpacking the Social Credit System

Imagine a society where trust is not just a virtue but a measurable, state-managed commodity. This is the ambitious, and often misunderstood, vision behind China's Social Credit System (SCS). It's an intricate web of technology and policy designed to build what the government calls a more "sincere" society. In the West, it's frequently painted with a broad, dystopian brush—a single, all-knowing score that dictates every citizen's fate. The reality, however, is a far more complex tapestry woven from big data, artificial intelligence, and deep-seated political philosophies.

This system isn't just about watching people; it's a paradigm for a new kind of social governance powered by AI. It functions on two primary levels: one aimed at corporations, ensuring they comply with regulations, and another aimed at individuals, subtly shaping their behavior to align with state-endorsed norms. To truly understand it, we have to move past the simple narratives and look at its architecture, its real-world applications, and the ancient ideas that give it life. The system was born from a specific concern—a perceived crisis of trust in the market after decades of breakneck economic growth. The core mechanisms it employs are not subtle; they are systems of blacklists for punishment and "redlists" for rewards. By exploring how it actually works, we can begin to have a more nuanced conversation about its profound implications for privacy, personal freedom, and the very nature of social control in our increasingly digital world.

The Philosophical Blueprint: Old Ideas in New Code

To grasp the logic of the Social Credit System, you have to look back centuries before the first line of code was ever written. It's a modern tool, but it's powered by very old philosophical engines. The first is a simple concept that any government might use: Utilitarianism. This is the idea that the best action is the one that brings the greatest good to the greatest number of people. The Chinese government frames the SCS in precisely these terms. They argue that by creating a system that ensures people and companies honor their contracts, follow the law, and act in good faith, society as a whole becomes more stable, predictable, and prosperous. From this perspective, the loss of some individual privacy is a justifiable price to pay for the collective benefit of widespread social trust.

The second, more critical, view comes from the work of the French philosopher Michel Foucault and his concept of the "Panopticon." The Panopticon was a prison design where a single guard could see any prisoner at any time, but the prisoners never knew if they were being watched. The result? Prisoners began to police their own behavior, internalizing the guard's gaze. China's SCS is a digital panopticon. With cameras on every corner and data trails from every transaction, citizens understand the possibility of being watched is constant. This awareness is meant to encourage self-censorship and conformity, making people their own regulators.

But these Western concepts are layered on top of a distinctly Chinese foundation. The system blends the ancient philosophy of Legalism, which argued for governance through strict, clear laws and harsh punishments, with the ideals of Confucianism, which emphasizes social harmony, moral integrity, and sincerity. The SCS's use of public blacklists and stark consequences is purely Legalist. At the same time, its stated goal of fixing a "moral vacuum" and building a culture of trust speaks directly to Confucian values. AI and big data didn't create this philosophy; they simply provided the means to implement these centuries-old ideas on a scale and with an efficiency that was previously unimaginable.

Deconstructing the "Score": The Myth vs. The Reality

One of the biggest myths about the Social Credit System is the idea of a single, universal "citizen score" that follows every person in China, rising and falling with every action. The image is powerful—buy too many video games, your score drops; donate to charity, it goes up. But as of 2025, this sci-fi vision is not the reality. The system is not one monolithic entity but a fragmented ecosystem of different programs. Think of it less as a single building and more as a collection of interconnected, and sometimes disconnected, databases and pilot projects.

Some cities, like Rongcheng, did experiment with a points-based system. In that pilot, citizens started with 1,000 points and could gain or lose them based on their behavior, striving for a high rating. However, this local model was never rolled out across the entire country. The national-level system, which is the most consequential, works very differently. It focuses less on a granular score and more on a blunt instrument: blacklists and redlists.

The most famous of these is the Supreme People's Court's blacklist for "judgment defaulters," colloquially known as laolai. This list is for individuals and companies who refuse to comply with a court's judgment, most often by failing to pay their debts. Getting on this list has severe and immediate consequences. It's not about a small deduction of points; it's about being fundamentally restricted. So, while online censorship certainly exists in

China, posting an unpopular opinion on social media is very unlikely to land you on the national laolai blacklist. Failing to pay a court-ordered fine, however, almost certainly will. The national system, therefore, is primarily a tool for enforcing legal and financial compliance, not for micromanaging everyday social etiquette.

The Technological Engine: How It All Works

The Social Credit System runs on a colossal engine of data collection and analysis. This technological backbone is what makes its scale and reach possible. The raw fuel for this engine is data, harvested from a vast array of sources. China's rapidly expanding network of surveillance cameras, part of projects like "Skynet" and "Sharp Eyes," plays a huge role. These aren't just simple cameras; many are equipped with advanced facial recognition, capable of identifying individuals in a crowd with stunning accuracy.

This visual data is then combined with a flood of other information. Government agencies contribute administrative records: everything from traffic violations and tax payments to court filings and business licenses. State-owned banks and financial institutions provide credit histories. Furthermore, data from China's tech giants can also be incorporated, although the exact mechanisms are often opaque. Information from online shopping habits, social media activity, and payment apps can all potentially become part of an individual's or company's file.

The role of Artificial Intelligence is not to act as a judge, but as an impossibly fast and powerful analyst. AI algorithms sift through these mountains of data to identify patterns, connect disparate pieces of information, and flag individuals or companies for specific behaviors. For instance, an AI might cross-reference court records with airline ticketing systems to automatically prevent a blacklisted individual from purchasing a plane ticket. It can fuse data from tax offices and customs to identify a company that might be engaged in fraudulent activity. The ultimate goal is integration—to create a unified, comprehensive profile for every entity, allowing for seamless, automated enforcement of rules and regulations.

Real-World Consequences and Ethical Dilemmas

For those who find themselves on the wrong side of the system, the consequences are far from theoretical. Being placed on a blacklist like the laolai list can dramatically shrink your world. Punishments are designed to be both inconvenient and publicly humiliating. A blacklisted person can be barred from purchasing high-speed train or plane tickets, effectively limiting their travel. They can be prevented from buying luxury goods, staying in nice hotels, or even enrolling their children in expensive private schools. In some regions, the faces and names of defaulters are displayed on public billboards and even before movies in cinemas. This isn't just a penalty; it's a form of public shaming designed to compel compliance.

The system's reach extends to the corporate world as well. Companies receive ratings based on their adherence to a wide range of regulations, including tax compliance, environmental standards, and product safety. A good rating—being on a "redlist"—can lead to benefits like lower tax rates, better loan terms, and fewer inspections. A poor rating, however, can trigger significant penalties, including exclusion from government contracts, higher inspection frequencies, and restricted access to credit. It creates a powerful incentive for businesses to stay in the government's good graces.

This raises profound ethical questions. Proponents argue it builds a high-trust society and efficiently enforces the rule of law. But critics point to the immense potential for error and abuse in such an opaque, centralized system. Where is the due process? How does one appeal a negative rating generated by a faceless algorithm? The constant surveillance creates a palpable chilling effect, discouraging not just illegal acts but also dissent, creativity, and any behavior that deviates from the norm. It forces a difficult conversation about the balance between the collective good and individual liberty, and whether the quest for social harmony is worth the price of personal autonomy.

The Corporate Scorecard: A Tool for Taming the Market

While stories of individual scores often capture the global imagination, the less-seen but arguably more powerful engine of the Social Credit System is the one aimed at businesses. The Corporate Social Credit System (CSCS) is the more mature, fully implemented sibling of the individual system. Think of it as a massive, digital report card for every single company operating in China, from the smallest local workshop to the largest multinational corporation. It's a vast regulatory database that is constantly listening, watching, and, most importantly, grading.

The system was born out of a real need. After decades of explosive, often chaotic economic growth, the government needed a more effective way to enforce its own rules. The CSCS became the answer. It pulls in data from dozens of government departments, creating a comprehensive profile of a company's behavior. Is the company paying its taxes on time? Is it following environmental laws? Are its products safe? Are its workplace safety records clean? Every compliance detail, big or small, feeds into the system.

The logic behind it is a simple but powerful carrot-and-stick approach. Companies that play by the rules and maintain a good record are put on "redlists." This is the reward. It can mean easier access to bank loans, fewer government inspections to deal with, and a smoother path through bureaucratic red tape. On the other hand, companies with poor records are slapped onto blacklists. This is the punishment. Suddenly, they might find themselves barred from bidding on lucrative government projects, restricted from issuing bonds, and under the constant microscope of regulators. For example, a factory that repeatedly dumps pollutants into a river won't just face a fine; its bad CSCS rating could block its ability to get a loan for a new facility, making its very future uncertain. It's a direct and potent way to make corporate misbehavior economically painful.

The Mechanisms of Control: A World of Digital Walls

For those who end up on a blacklist, whether an individual or a corporation, the consequences are swift, tangible, and designed to be felt. The system operates on a principle the government calls "joint punishment." This is a powerful concept: a failure in one area of your life can trigger a cascade of restrictions across many others. It's a digital domino effect. Your world can suddenly shrink, surrounded by walls you can't see but that are very, very real.

The most famous of these punishments are the travel restrictions. Imagine needing to travel for work or to visit a sick relative, but when you try to buy a plane ticket or a high-speed train ticket, the system simply says no. You are grounded. This is a daily reality for millions of people on the judgment defaulter blacklist, often for failing to pay a court-ordered debt. It's an incredibly effective way to compel compliance by severely limiting personal and professional mobility.

But the walls don't stop at the train station. Penalties can cut you off from financial life, making it nearly impossible to get a loan or a credit card. They can lock you out of certain jobs, particularly in the government or state-owned enterprises. In some of the more extreme cases, a parent's bad record can even prevent their children from getting into top-tier private schools.

Perhaps the most potent tool is public shaming. This is not a subtle mechanism. The system leverages the deep cultural importance of social reputation, or "face." The names, photos, and partially redacted ID numbers of blacklisted people are often displayed on massive public screens in city squares, on buses, and even in movie theaters before the film starts. Your failure becomes public entertainment and a warning to others. This combination of material punishment and public humiliation creates immense pressure to resolve the issue that put you on the list in the first place.

The AI Engine: Connecting the Digital and Physical Worlds

The entire Social Credit System is built on a foundation of technology that was once the stuff of science fiction. Its backbone is the state's immense capacity to collect, process, and connect oceans of data. Artificial intelligence is the nervous system that makes this all possible, turning raw information into actionable control. At the heart of this is China's sprawling network of hundreds of millions of surveillance cameras. But these are not just passive recording devices; increasingly, they are smart eyes, powered by some of the world's most advanced facial recognition AI.

This network allows authorities to do more than just watch public spaces. It allows them to track individuals, linking a face seen at a protest or a crosswalk to a specific digital identity in a government database. The grand vision, though still a work in progress, is to achieve total data integration. The goal is to merge the data from this physical surveillance with everyone's digital footprint. This includes your financial history, your court records, your online shopping habits, and even your social media activity.

AI algorithms are the essential tool for making sense of this data deluge. They are designed to sift through these trillions of data points to identify patterns, assess risk, and automatically flag a person or a company for a blacklist. While the system is not yet a single, all-knowing entity, its ambition is clear: to build a complete, 360-degree digital profile of every citizen and business. The most extreme and controversial applications of this technology, such as the alleged use of AI in the Xinjiang region to detect emotional states, show the terrifying potential for this surveillance to move beyond punishing actions to preemptively controlling thoughts and dissent. It raises profound questions about whether a person can truly be free when their every move and expression is being cataloged and judged by an algorithm.

What It All Means: A Global Case Study

The Social Credit System is more than just a domestic policy; it's a globally significant event with far-reaching implications. For any international company operating in China, it has become an unavoidable reality. Managing your Corporate Social Credit profile is no longer an option;

it is a core part of business strategy and survival. A low score isn't just a mark on a report card; it can directly threaten your market access, disrupt your supply chain, and invite crushing regulatory scrutiny. Ignoring your CSCS score is like ignoring your taxes—it's a risk that can lead to severe consequences.

For the rest of the world, China is running the largest experiment in AI-powered social engineering in human history. It serves as a stark, real-world case study on how technology can be used to enforce state control on a massive scale. It presents a powerful and alternative vision of governance, one where order and stability are prioritized, even at the cost of individual privacy and autonomy. This model stands in sharp contrast to the rights-based, privacy-focused data governance frameworks slowly emerging in Europe and other parts of the world.

Ultimately, the Social Credit System forces us all to confront fundamental questions about the future we are building. How much privacy are we willing to trade for security and convenience? Can a system designed for mass control ever be truly fair, transparent, or accountable? As AI becomes more integrated into our lives, the lessons from China's experiment—both its stated successes and its chilling excesses—will become increasingly relevant for every society grappling with the power of technology.

United States - Algorithmic Bias in Criminal Justice

Theoretical Underpinnings of Algorithmic Bias

The introduction of algorithms into the United States criminal justice system is not merely a technological update; it is a profound ethical challenge that strikes at the heart of what we mean by "justice." These complex mathematical systems, designed to predict human behavior, force us to confront a fundamental tension in American legal philosophy: the clash between treating everyone the same and ensuring everyone gets a fair outcome. This debate is framed by two competing ideas of equality.

First, there is formal equality. This is the classic idea of justice being blind. It argues that the fairest system is one that applies the exact same rules and procedures to every single person, regardless of their background, race, or economic status. It focuses on the process, insisting that if the rules are identical for all, then justice has been served. It's about consistency.

On the other hand, there is substantive equality. This perspective argues that simply applying the same rules to everyone is not enough, because people don't start from the same place. It looks beyond the process and focuses on the results. Substantive equality is concerned with achieving equitable outcomes, which might require treating people differently to correct for historical and systemic disadvantages. It acknowledges that a playing field can be tilted, and true fairness sometimes means giving some people a boost to make the game genuinely equal.

The controversy surrounding algorithms like COMPAS is a perfect illustration of this conflict. The algorithm's designers defended their creation by pointing to a form of formal equality—its predictive accuracy was roughly the same for different racial groups. However, critics immediately pointed to its failure to achieve substantive equality, as the types of errors it made disproportionately harmed Black defendants.

This leads to another critical concept: procedural justice. This theory suggests that people's perception of fairness is tied not just to the final outcome, but to the transparency and impartiality of the process itself. For a system to be seen as just, people must understand how decisions are made and believe that those decisions are unbiased. Proprietary, closed-source algorithms present a direct challenge to this principle. They operate as "black boxes." We can see the data that goes in and the risk score that comes out, but the internal logic—the very reasoning behind a decision that could alter someone's life—is hidden from view, known only to the private company that owns it. This lack of transparency makes it nearly impossible to question or appeal a decision, undermining the very essence of due process.

Ultimately, the fierce debate over these algorithms is about much more than code. The AI system is not inventing bias from scratch. Instead, it is a mirror reflecting the deeply embedded biases that already exist in our society. It is fed historical data on arrests, charges, and convictions—data that is already shaped by generations of systemic racism, from disproportionate policing in minority neighborhoods to socioeconomic disparities that can lead to higher crime rates. The algorithm learns from our biased past and projects that bias into the future, but it does so with a veneer of objective, mathematical certainty. In doing so, it forces a direct, quantifiable confrontation with the inequalities America has long struggled with. The fight for algorithmic fairness is simply the next chapter in the long-standing civil rights struggle, now being waged on the complex new battlefield of data science.

The COMPAS Case Study: Unpacking ProPublica's Investigation

In May of 2016, the investigative journalism outlet ProPublica published a report that sent shockwaves through the legal and tech communities. Titled "Machine Bias," the investigation became a foundational text for understanding the real-world dangers of algorithmic bias. The report was not a theoretical exercise; it was a deep, data-driven analysis of a tool being used to make life-altering decisions about thousands of people across the United States.

The algorithm at the center of the storm was the Correctional Offender Management Profiling for Alternative Sanctions, or COMPAS. Developed by a private company called Northpointe (now Equivant), COMPAS was a risk assessment tool used in courtrooms and parole offices nationwide. It analyzed a defendant's history and personal information to generate a score from 1 to 10, predicting their likelihood of committing another crime. This score could influence everything from whether a person was granted bail to the length of their prison sentence. The promise was simple: to bring data-driven objectivity to a messy, human process.

ProPublica's investigation put that promise to the test. Their team obtained the COMPAS risk scores for more than 7,000 people arrested in Broward County, Florida, over a two-year period. They then spent months meticulously checking the algorithm's predictions against the actual outcomes—did the people flagged as high-risk actually go on to re-offend?

Their findings were stark and deeply troubling. The algorithm was not the neutral arbiter it claimed to be; it was systematically biased against Black defendants. The data revealed several key disparities:

The algorithm was far more likely to falsely label Black defendants as high-risk for future crime. ProPublica found that Black defendants were almost twice as likely as white defendants to be incorrectly flagged as high-risk when they, in fact, did not re-offend.

The false positive rate for Black defendants was a staggering 45%. This means that of the Black defendants who would not commit another crime, nearly half were misclassified by COMPAS as being a high risk to do so. The same false positive rate for white defendants was only 23%.

The system also made the opposite mistake, but in a way that favored white defendants. White defendants who did go on to commit new crimes were mislabeled as low-risk much more often than Black defendants who re-offended. The false negative rate for white defendants was 48%, compared to just 28% for Black defendants.

To make these statistics devastatingly clear, ProPublica highlighted the human stories behind the numbers. They contrasted the case of Brisha Borden, a Black 18-year-old who was rated as a high risk, with Vernon Prater, a white 41-year-old rated as a low risk. Borden and a friend had playfully taken a child's bicycle and scooter, a minor offense. Prater was an experienced armed robber. The algorithm's predictions were completely backward. Borden never re-offended. Prater later committed a serious burglary and is now serving an eight-year prison sentence. These powerful examples showed that the algorithm's biased errors weren't just numbers on a page; they were injustices with profound and lasting consequences for real people.

The Inherent Trade-off: Competing Definitions of Algorithmic Fairness

The response to ProPublica's investigation was as revealing as the report itself. Northpointe, the creator of COMPAS, did not dispute the numbers. They acknowledged the disparity in error rates that ProPublica had uncovered. However, they argued that their algorithm was not biased. Their defense hinged on a critical and often misunderstood point: in the world of data science, "fairness" does not have one single, universally accepted definition.

This controversy brought a complex technical debate out of academic circles and into the public square. It turns out that when the underlying data shows different base rates of an outcome for different groups—in this case, different recidivism rates for Black and white defendants—it is mathematically impossible for a risk assessment tool to satisfy all major definitions of fairness at the same time. You are forced to choose which definition of fairness you want to prioritize, and that choice is a reflection of your values.

The COMPAS debate was a battle between two competing fairness metrics:

Error Rate Balance: This is the definition ProPublica used to measure fairness. Also known as "equalized odds," this metric demands that the algorithm's mistakes are distributed equally across all groups. It requires that the false positive rate (wrongly flagging someone as high-risk) and the false negative rate (wrongly flagging someone as low-risk) are the same for both Black and white defendants. The core value here is protecting individuals from the harm of being misclassified. ProPublica's data clearly showed that COMPAS failed this test. The burden of being wrongly labeled a future criminal fell much more heavily on Black individuals.

Predictive Parity: This is the definition Northpointe used to defend its algorithm. Also known as "calibration," this metric requires that a risk score means the same thing regardless of a person's race. If the algorithm assigns a score of "7," it should mean that individuals in that group have, for example, a 70% chance of re-offending, whether they are Black or white. The core value here is ensuring the statistical accuracy and reliability of the prediction itself. Northpointe presented data showing that COMPAS did satisfy this condition. The algorithm's overall accuracy in predicting who would re-offend was roughly equal for both groups, at around 61%.

This conflict reveals that "fairness" is not a neutral, technical property that can be programmed into a machine. The choice of which metric to prioritize is a normative decision—a moral judgment. Do we design a system that prioritizes equalizing the harm of its inevitable mistakes, even if it means the scores are less statistically consistent across groups? Or do we prioritize the raw predictive accuracy of the scores, even if it means the mistakes will disproportionately harm one group over another?

The COMPAS algorithm, and the brilliant investigation into its flaws, did not provide an easy answer. Instead, it forced a crucial and uncomfortable realization. By embedding our data into these systems, we are embedding our societal values—and our societal biases—as well. Algorithms don't eliminate the need for difficult human choices; they simply hide those choices behind a wall of code and statistics, forcing us to decide what kind of fairness we truly value most.

Sources of Bias: From Data to Design

When we find racial disparities in the outcomes of an algorithm like COMPAS, it's natural to imagine a programmer intentionally writing biased code. But the reality is far more complex and insidious. The bias doesn't typically come from a single line of malicious code. Instead, it seeps into the system through multiple channels, starting long before the first line of code is even written. It is a problem born from the data we feed it and the human choices we make in designing it.

The single most significant source of bias is the data itself. Algorithms learn from the information they are given, and in the United States, the historical data of our criminal justice system is a reflection of our society's long and troubled history with race. For decades, policing strategies, such as "broken windows" or "stop and frisk," have disproportionately targeted minority communities. This leads to higher arrest rates for minor offenses in these neighborhoods, not necessarily because more crime is happening, but because more police are present to witness and record it. For example, studies have consistently shown that while white and Black Americans use illegal drugs at similar rates, Black individuals are arrested for drug possession at a significantly higher rate. When an algorithm is trained on this data, it doesn't see the systemic injustice. It only sees a statistical pattern: people from certain neighborhoods, who are predominantly people of color, are arrested more often. The algorithm learns this pattern and concludes that these individuals are inherently higher risk, thus building historical discrimination into its predictive model and projecting it into the future.

Furthermore, while these algorithms are legally prohibited from using race as a direct input, they achieve a similar result by using proxy variables. These are factors that are not explicitly race but are so closely correlated with it that they act as stand-ins. The most common proxy is a person's zip code. Due to a long history of residential segregation and economic inequality, a person's neighborhood is often a strong indicator of their race and socioeconomic status. The COMPAS questionnaire digs even deeper, asking questions that are heavily influenced by a person's environment, such as whether their parents separated, if they have friends who have been arrested, or if they live in a "crime-ridden" neighborhood. For a person growing up in a heavily policed, under-resourced community, these factors are often beyond their control. Yet the algorithm interprets these environmental disadvantages as indicators of individual risk, effectively penalizing people for the circumstances into which they were born.

Finally, human choices made during the algorithm's design can introduce or amplify bias. The developers decide which variables to include in the model and how much importance, or "weight," to assign to each one. Is a prior arrest for trespassing as significant as a prior arrest for assault? Should a person's employment status be a major factor? These are not objective, mathematical decisions; they are value judgments. These choices, even if made with the best intentions, can reflect the unconscious biases of the development team, further skewing the algorithm's output in ways that can be difficult to detect or challenge once the system is built. The bias, therefore, is not an accident but an inevitable consequence of building a supposedly neutral system on a foundation of unequal data and subjective human decisions.

The Regulatory and Judicial Response

The United States' effort to grapple with the rise of AI in its justice system has been a scattered and evolving affair. Unlike the European Union, which has moved toward a comprehensive, top-down legal framework like the AI Act, the American response has been a patchwork quilt. It is a mix of high-level guidance from the federal government, specific and often restrictive laws at the state and local levels, and critical legal battles being fought in the courtroom. The common thread running through this tangled web is a growing demand for three things: transparency, auditing, and unwavering human oversight.

At the federal level, the response has been more about setting principles than writing hard-and-fast rules. The Department of Justice and the White House have issued executive orders and governance frameworks that call for the "responsible and ethical" use of AI. These documents emphasize the need to protect civil rights and civil liberties, promote public trust, and ensure that AI systems are fair and equitable. While important for setting a national tone, this guidance often lacks the teeth of direct regulation, leaving much of the concrete policymaking to other levels of government.

It is at the state and local levels where the most significant battles are being waged. A wave of legislation has emerged, particularly focused on the most controversial AI technologies like facial recognition. Cities like San Francisco, Oakland, and Boston have passed outright bans on the use of facial recognition technology by police and other municipal agencies. Other states have passed laws requiring law enforcement to obtain a warrant before using such tools. These actions represent a growing public skepticism and a desire to pump the brakes on the adoption of powerful surveillance technologies until their fairness, accuracy, and constitutional implications can be thoroughly understood.

However, the greatest single obstacle to accountability remains the "black box" problem. Most risk assessment tools, including COMPAS, are created by private companies that guard their algorithms as valuable trade secrets. This means that defense attorneys, defendants, and even judges are prevented from seeing the exact calculations and logic that produced a risk score. This is not merely a technical inconvenience; it represents a fundamental challenge to the constitutional right of due process. How can a defendant confront the evidence against them when that evidence is a secret formula locked in a corporate vault?

In response, the judicial system and policymakers are increasingly pushing back, not by trying to break open the black boxes, but by limiting their power. The emerging consensus is that an algorithm's output can only ever be a piece of information, not a final verdict. Courts are reinforcing the principle that a human judge must retain ultimate decision-making authority. This judge is legally and ethically required to consider all the unique facts of a case, including arguments from the defense that highlight a person's individual circumstances—information that a one-size-fits-all algorithm could never capture. The goal is to ensure that AI remains a tool to inform, not a command to be obeyed, preserving the essential role of human discretion in the delicate process of administering justice.

Practical Implications

For the judges, lawyers, police departments, and policymakers working on the front lines of the American justice system, the story of COMPAS is not an abstract academic debate. It is a profound and urgent cautionary tale. It demonstrates with startling clarity that adopting algorithmic tools without deep skepticism and rigorous oversight is not a step toward a more objective future, but a dangerous leap that risks automating and entrenching the very biases we are striving to overcome. The key takeaway is that these systems cannot be trusted blindly; they must be interrogated, audited, and held accountable at every step.

The first and most critical implication is the need for exhaustive due diligence before any such tool is ever deployed. Law enforcement agencies and court systems must shift from being passive consumers to being aggressive investigators. This means demanding complete transparency from the private vendors selling these algorithmic products. Stakeholders must ask the hard questions: What data was used to train this model? What are the known limitations and error rates for different demographic groups? How have you tested for bias? A vendor's refusal to provide clear answers should be an immediate red flag. The era of accepting proprietary "black box" systems on faith must end.

Second, independent bias audits should become a non-negotiable prerequisite for the use of any risk assessment tool. These audits cannot be conducted by the vendor themselves; they must be performed by neutral, third-party experts who can test the algorithm's performance in the real world, using local data. Crucially, these audits must evaluate the tool against multiple definitions of fairness. It is no longer acceptable for a company to claim its product is "fair" based on a single metric like predictive parity while ignoring devastating disparities in error rates. A comprehensive audit would expose these trade-offs, allowing communities to have an informed public debate about which definition of fairness they value most and what level of error they are willing to accept.

Most importantly, we must recognize that technology can never be a replacement for human judgment. The ultimate backstop against algorithmic bias is an empowered and well-trained human decision-maker. It is not enough to simply keep a "human in the loop"; that person must be equipped to critically assess the algorithm's recommendation and have the confidence to override it when necessary. This requires training judges and parole officers not just on how to read a risk score, but on the inherent limitations of these tools—how they can be skewed by biased data and how they often fail to capture an individual's capacity for growth and change. In the final analysis, the pursuit of justice is not a statistical problem to be solved. It is a deeply human endeavor that requires empathy, context, and wisdom—qualities that can never be programmed into a machine.

Part II: AI in High-Stakes Societal Domains

United Kingdom - AI in Healthcare: Innovation vs. Patient Rights

Abstract

This chapter utilizes the United Kingdom's National Health Service (NHS), one of the world's largest single-payer healthcare systems, as a case study to explore the immense potential and significant ethical challenges of integrating artificial intelligence into healthcare at a national scale. It examines the powerful drive for innovation to improve diagnostics, personalize treatment, and enhance operational efficiency, which promises to transform patient outcomes. This ambition is set against a backdrop of critical ethical and legal concerns, including the privacy of patient data, the transparency and accountability of "black box" algorithms, the potential for liability in cases of AI-driven error, and the imperative to ensure equitable access to the benefits of these new technologies. The UK's approach, characterized by strategic investment, centralized data initiatives, and the development of novel regulatory sandboxes, provides a compelling model for navigating this complex landscape.

Theoretical Underpinnings: The Ethical Compass of AI in Medicine

At the heart of the debate over AI in healthcare lie four foundational principles of Bioethics. These principles aren't just academic concepts; they are the moral compass guiding every decision, from designing an algorithm to deploying it in a hospital ward. They provide a sturdy framework for asking the hard questions about what it means to innovate responsibly.

The first principle is Beneficence, which is the simple, powerful obligation to do good. In the world of the NHS, this means that any AI tool must be developed with the clear goal of making patients healthier and improving their lives. It's not about technology for technology's sake. It's about creating systems that can predict a patient's risk of developing sepsis, identify early-stage cancers with greater accuracy, or personalize a treatment plan for diabetes. The ultimate test for any healthcare AI is whether it produces a tangible, positive outcome for the people the NHS serves.

Conversely, we have Non-maleficence, the duty to "do no harm." This principle forces us to confront the risks. An AI system is only as good as the data it's trained on. If that data reflects historical biases—for example, if it's primarily from one demographic—the algorithm could perform poorly for other groups, leading to misdiagnoses and genuine harm. A data breach that exposes sensitive patient records is another form of harm. This principle demands a relentless, clear-eyed assessment of every potential danger before a system is allowed anywhere near a patient.

Autonomy champions the right of every patient to make their own decisions about their body and their care. This becomes complicated with AI. True informed consent means a patient must understand what they are agreeing to. But how can a patient give meaningful consent to the use of a complex algorithm they can't possibly understand? How do we explain that their data will be used to train a machine? Respecting autonomy in the age of AI means finding new and better ways to communicate, ensuring that patients remain the ultimate decision-makers in their healthcare journey, not passive subjects of a machine's logic.

Finally, the principle of Justice demands fairness. It insists that the incredible benefits of AI shouldn't just go to the wealthy or the tech-savvy. In the context of a national health service, this is paramount. Will a cutting-edge diagnostic tool be available in a rural clinic in Cornwall as well as a major London hospital? Will it work equally well for people of all ethnicities and backgrounds? Justice forces us to consider how we distribute these powerful new tools so they close existing health gaps, rather than widening them. The UK's centralized NHS structure makes this a high-stakes issue. Its vast, unified dataset is a goldmine for training powerful AI. But this same centralization means a single biased algorithm deployed nationwide could cause harm on an unprecedented scale, shaking public trust in the very institution designed to protect them. This explains the UK's careful dance: moving forward with ambition while building strong guardrails like national commissions and regulatory testbeds to manage the risks.

The Drive for Innovation: AI for a Modern NHS

The NHS is not just experimenting with artificial intelligence; it is actively embedding it into its long-term vision. Faced with an aging population, rising costs, and workforce pressures, the UK government and NHS leaders see AI as a critical tool for building a more efficient, effective, and sustainable healthcare system. This isn't a distant future—it's happening right now, backed by significant government investment and a clear strategic direction to harness AI's power to prevent, diagnose, and treat disease.

One of the most dramatic impacts is in medical imaging. Radiologists and other clinicians are tasked with analyzing thousands of scans, a process that is both time-consuming and requires immense concentration. AI is emerging as an invaluable "second pair of eyes." These algorithms can sift through CT scans, MRIs, and X-rays, flagging subtle anomalies that might escape the human eye. They can prioritize the most urgent cases, ensuring that patients who need immediate attention get it faster. This doesn't replace the expert clinician; it augments their ability, freeing them from repetitive tasks to focus on complex cases and direct patient interaction.

Beyond spotting disease, AI is also being deployed to streamline the very way hospitals run. It's being used to predict patient admission surges, allowing hospitals to better manage bed capacity and staffing levels. Administrative tasks that consume countless hours of clinical time, like transcribing notes and scheduling appointments, are being automated. Each minute an AI saves a doctor or nurse on paperwork is another minute they can spend at a patient's bedside.

The real-world examples are compelling and growing in number:

Example 1: Diagnostics in Action. In a remarkable national rollout, AI is now standard practice in every single stroke unit across England. It analyzes brain scans of suspected stroke patients with incredible speed, helping doctors make faster, more precise decisions about the best course of treatment, where every second saved can mean the difference between recovery and permanent disability. In another powerful example, seven hospital trusts in the North East are using an AI platform to review chest X-rays. This system has been shown to improve the accuracy of lung cancer detection by a staggering 45%, catching the disease earlier when it is most treatable.

Example 2: Efficiency and Patient Comfort. The NHS has embraced a technology called HeartFlow. Instead of requiring an invasive angiogram to diagnose coronary heart disease, this technology uses AI to analyze a standard CT scan. From this scan, it builds a personalized, color-coded 3D model of the patient's coronary arteries, showing clinicians exactly where blockages are restricting blood flow. This AI-driven approach is not only less risky and more comfortable for the patient, but it also cuts diagnostic costs by around 25%, delivering better care for less money.

Ethical Hurdles: Data Privacy and the 'Black Box' Problem

The fuel for any healthcare AI is data—vast amounts of patient data. This reality immediately brings us to one of the most significant ethical minefields: privacy. The information held by the NHS is among the most personal and sensitive data imaginable, detailing every diagnosis, treatment, and vulnerability of millions of citizens. The public's trust that this data will be protected is the bedrock of the entire system. UK law, including the stringent General Data Protection Regulation (GDPR), imposes strict rules on how this data can be used. Techniques like "anonymization" or "pseudonymization" are used to strip personal identifiers from datasets before they are used to train AI models. However, these methods are not foolproof, and there is an ongoing debate about whether true anonymization is even possible with complex health data. High-profile controversies, such as the data-sharing agreement between Google's DeepMind and the Royal Free Hospital in London, have highlighted public anxiety and underscored the need for absolute transparency in how patient data is used.

Beyond privacy is the challenge of the "black box." Many of the most powerful AI systems, particularly those using deep learning, operate in a way that is not easily understood by humans. We can see the data that goes in (the patient's scan) and the result that comes out (the diagnosis), but the intricate, multi-layered reasoning process in the middle is often opaque. This is a profound problem in medicine. A doctor cannot simply trust a machine's output without question. They are ethically and legally responsible for their patient's care. If an AI recommends a radical treatment, the doctor needs to understand why. They need to be able to explain that reasoning to their patient. Without this transparency, accountability breaks down, and the trust between doctor and patient is jeopardized. This has spurred the development of a new field called "Explainable AI" (XAI), which aims to create systems that can justify their conclusions in a way that humans can understand, turning the black box into a more transparent glass box.

Governance and Regulation: Navigating Liability and Equity

As AI tools become more integrated into clinical practice, they raise thorny questions that stretch our existing legal and regulatory frameworks to their limits. Perhaps the most pressing question is about liability: if an AI-driven diagnosis is wrong and a patient is harmed, who is at fault? Is it the doctor who accepted the AI's recommendation? Is it the hospital that purchased and deployed the software? Or is it the company that developed the algorithm? There are no easy answers, and the law is still racing to catch up with the technology. Currently, the responsibility largely remains with the clinician, who is expected to use AI as a tool and apply their own professional judgment. However, this creates immense pressure. To manage this, regulatory bodies like the UK's Medicines and Healthcare products Regulatory Agency (MHRA) are developing new frameworks to certify and monitor AI as a medical device, ensuring it meets rigorous standards of safety and effectiveness before it can be used.

The principle of justice also demands a robust governance structure to ensure equity. The risk of creating a two-tiered health system—one for those in affluent, well-funded areas with access to the latest AI diagnostics, and another for everyone else—is very real. The NHS, as a national service, has a unique obligation to ensure that the benefits of innovation are distributed fairly across the entire population, regardless of geography or socioeconomic status. This means consciously planning the rollout of new technologies, investing in digital infrastructure in underserved areas, and ensuring that algorithms are trained on diverse datasets so they work for everyone.

To navigate this complex intersection of innovation and risk, the UK has pioneered the use of "regulatory sandboxes." This is a clever approach that creates a safe, controlled environment for companies to test new AI technologies on real NHS data under the close supervision of regulators. It allows innovation to happen at pace, but with guardrails in place. Companies can see how their technology performs in the real world, and regulators can learn about the risks and develop appropriate rules in tandem. It is a pragmatic and collaborative model, designed to foster a world-leading AI ecosystem while never losing sight of the primary duty to protect patients. This balanced approach—combining ambitious national strategy with cautious, iterative governance—is the UK's signature contribution to the global challenge of building a future where AI in healthcare is not only powerful but also safe, ethical, and just for all.

Data Privacy and Governance: The NHS's Digital Crown Jewels

The National Health Service (NHS) isn't just a healthcare provider; it's the guardian of one of the world's most extraordinary collections of human health data. Imagine a library containing the detailed, lifelong medical stories of over 65 million people. This isn't just a random collection of facts; it's a longitudinal treasure trove, tracking individuals from birth to old age, through every illness, recovery, and treatment. For scientists and AI developers, this data is the ultimate fuel. It's the raw material needed to train artificial intelligence that can predict disease outbreaks, spot cancer earlier than the human eye, and tailor treatments to an individual's unique genetic makeup. The potential to revolutionize medicine and save countless lives is immense.

However, this incredible power creates a profound dilemma. Every single data point in that collection represents a person—their vulnerabilities, their struggles, and their most private health information. How can the UK unlock the immense public good hidden within this data without betraying the fundamental right to privacy for every citizen? This is the tightrope the NHS must walk. Public trust is the safety net, and if it tears, the entire enterprise could fall. To manage this, the UK's data protection laws, which echo the stringent EU General Data Protection Regulation (GDPR), impose very strict rules. Health data is rightly classified as "special category data," meaning it requires the highest level of protection.

To solve this puzzle, the NHS has pioneered the concept of a Secure Data Environment (SDE). Think of it not as a data giveaway, but as a highly secure reading room in a library. Approved researchers are granted access to a virtual space containing vast pools of de-identified data—where names, addresses, and other direct identifiers have been stripped away. They can bring their algorithms and analytical tools into this environment to work on the data. The crucial rule is that the raw data itself can never be removed. It's a "data-in, insights-out" model. This allows for powerful research while ensuring the foundational information remains under the strict control and governance of the NHS. The new NHS Federated Data Platform (FDP) extends this idea, aiming to connect data across different NHS trusts to improve operational efficiency—like managing patient flow and bed capacity—without pooling all the data into one central location, further enhancing security and control. These measures are designed to build a fortress around patient information, fostering a system where innovation can flourish without sacrificing the privacy that underpins the public's relationship with the NHS.

Accountability and Liability: Who's Responsible When the AI is Wrong?

The rise of AI in medicine introduces a perplexing new challenge often called the "black box" problem. Imagine a highly sophisticated AI diagnostic tool that analyzes a patient's scans and correctly identifies a rare disease 99% of the time—a far better rate than any human doctor. But when asked how it reached its conclusion, the AI cannot explain its reasoning. Its internal decision-making process is a complex web of algorithms so intricate that even its own creators don't fully understand the "why" behind each specific answer. This lack of transparency creates a terrifying gap in accountability. If that AI makes a mistake in the remaining 1% of cases, and a patient is harmed as a result, who is to blame?

This question sends ripples through the entire healthcare system, creating a complex legal and ethical maze. Is the software company that developed the algorithm at fault for a flaw in its code? Or is it the hospital that purchased and deployed the technology, thereby endorsing its use? Perhaps the responsibility lies with the human clinician who ultimately accepted the AI's recommendation and acted upon it. Without a clear chain of responsibility, there is a risk that victims of an AI error could be left with no clear path to justice. This isn't just a theoretical problem; it's a fundamental barrier to the widespread and confident adoption of AI in high-stakes medical decisions.

To untangle this knot, UK regulators are moving away from old frameworks designed for static medical devices like scalpels or pacemakers. An AI is different; some advanced systems are designed to learn and adapt over time, meaning their performance can change after they've been deployed. The Medicines and Healthcare products Regulatory Agency (MHRA) is at the forefront of crafting a new rulebook for this new era. A key initiative is the "AI Airlock," a regulatory sandbox that acts as a controlled testing ground. It allows innovative AI technologies to be used in a real but supervised clinical setting. Think of it as a probationary period where regulators can watch how the AI performs, how it learns, and what its potential failure points are before it's approved for nationwide use in the NHS. This hands-on approach allows the MHRA to understand the unique risks of adaptive algorithms and to build a regulatory framework that is flexible enough to handle this dynamic technology, ensuring that when AI is integrated into patient care, there are clear lines of accountability and robust safety nets in place.

Equity and Access: Building an AI for Everyone

Artificial intelligence holds the promise of being a great equalizer in healthcare, offering expert-level diagnostics and personalized care to everyone, regardless of where they live. However, it also carries a sinister risk: the potential to amplify and entrench the very health inequalities it has the power to solve. The core of this problem lies in the data used to train AI models. An AI is only as good, and as fair, as the information it learns from. If an algorithm is trained predominantly on health data from a single demographic group—for instance, white males of a certain age—it will naturally become an expert in diagnosing conditions for that specific group.

The consequences of this are stark. That same algorithm may be significantly less accurate when analyzing data from women, ethnic minorities, or people from different socioeconomic backgrounds. A well-documented example is in dermatology, where algorithms trained on light-skinned individuals often fail to correctly identify cancerous skin lesions on darker skin, leading to missed diagnoses and potentially tragic outcomes. This isn't malicious; it's a reflection of historical biases present in the data we collect. If we're not careful, we could build a future where AI provides state-of-the-art care for some, while delivering substandard or even dangerous advice to others.

This challenge extends beyond algorithmic bias to the issue of access. There is a real danger of creating a "postcode lottery" for digital health. Will cutting-edge AI tools be deployed only in major, well-funded university hospitals in large cities, while smaller clinics and rural communities are left behind with older, less effective systems? This would create a two-tiered system of healthcare, fundamentally undermining the NHS's core principle of providing universal and equitable care for all. Recognizing this threat, NHS England and the UK's national AI strategy have placed fairness at the heart of their mission. They are developing tools like the Algorithmic Impact Assessment, which forces developers and hospitals to proactively evaluate whether a new AI tool could disproportionately harm certain patient groups before it is deployed. The ultimate goal is not just to innovate but to innovate responsibly, ensuring that the incredible benefits of AI are used to close, not widen, the gaps in health outcomes across the diverse tapestry of British society.

Japan - AI and Robotics for an Aging Population

The Heart of the Machine: Japan's Human-Centric Vision for a Robotic Future

Imagine a world not where robots have taken over, but where they have been invited in. A world where they aren't replacements, but partners. This isn't the plot of a distant science fiction film; it's the reality unfolding across Japan. The nation stands at the forefront of a profound global shift, a place where the twin forces of a rapidly aging population and a shrinking workforce are not seen as an impending apocalypse, but as the primary catalyst for innovation. Japan is pioneering the integration of artificial intelligence and robotics into the very fabric of its society. This exploration isn't just about technological wizardry; it's a deeply human story about survival, dignity, and the quest for harmony.

We will journey into how these advanced technologies are being woven into the most critical sectors, from the quiet intimacy of elder care homes to the bustling floors of factories and service industries. This analysis moves beyond the simple, often fearful, narrative of automation. It reveals a far more nuanced and thoughtful approach, where technology serves as an essential tool for support and augmentation. It's about creating a sustainable future where human hands are helped, not made obsolete. By examining the unique cultural philosophies, economic necessities, and ethical questions that shape this emerging human-robot coexistence, we can understand why Japan serves as a crucial, living laboratory for any nation staring down the barrel of a similar demographic future.

Theoretical Underpinnings: A Philosophy of Coexistence

To truly grasp Japan's journey with robotics, we have to look beyond the circuit boards and programming. The country's approach is built on a rich foundation of philosophical and cultural ideas that prioritize people first. It's a vision where technology must adapt to society, not the other way around.

This entire movement is best understood through the framework of Sociotechnical Systems Theory. It sounds complicated, but the idea is simple and profound: technology and society are not separate things. They are intertwined, like two dancers in a complex performance, each responding to and shaping the other's movements. In Japan, the development of a new robot isn't just a technical challenge for engineers in a lab. It is a direct societal answer to a demographic crisis. A shrinking workforce is a social problem; a collaborative robot that helps a smaller team assemble cars is the technical answer. In turn, the introduction of that robot changes how the team works, how they communicate, and what skills they value. This mutual shaping is constant. For instance, early elder care robots were often designed based on engineering principles alone, but through trials with actual seniors and their caregivers, they were redesigned to be less intimidating, with softer materials and more intuitive controls, reflecting a direct response to human social and emotional needs.

Moving from the societal level to the deeply personal, the ethical questions are viewed through the lens of Care Ethics. This perspective pushes aside rigid rules and instead focuses on the importance of relationships, connection, and the moral value of providing care. It asks the tough questions that get to the heart of what it means to be human. Can a robot, no matter how advanced, truly fulfill the relational part of care? Or is its role destined to be purely functional? A robot can lift a patient, deliver medicine on time, and monitor vital signs with perfect accuracy. This is the functional aspect of "caring for" someone. But can it provide the genuine emotional support, the empathy, and the shared connection that comes from "caring about" someone?

The Japanese answer to this dilemma is not to try and build a perfect artificial human. Instead, it is to design technology that supports and enhances the human element of care. A powerful robotic exoskeleton doesn't replace a caregiver; it gives that caregiver the strength to lift a loved one without injury, allowing them to save their energy for conversation, for holding a hand, for simply being present. In this model, the robot handles the strain, so the human can handle the connection.

This all flows from a foundational cultural philosophy: augmentation rather than replacement. This isn't just a practical response to labor shortages; it's rooted in the deep-seated societal value of wa (和), which translates to harmony and social balance. The Western narrative around AI is often one of conflict—human versus machine, a battle for jobs and relevance. The Japanese narrative is fundamentally about finding a harmonious coexistence. This leads to the development of technologies designed specifically to work alongside people. You see this in "cobots" (collaborative robots) on assembly lines, which are designed to be safe enough to work right next to humans, handling repetitive or strenuous tasks while the human worker performs tasks requiring dexterity and critical thinking.

This philosophy is so central that it has been codified in government-backed principles like the "Social Principles of Human-Centric AI." These guidelines explicitly state that AI should not infringe on human dignity and autonomy. The goal is to balance incredible technological progress with the preservation of social harmony. The results are wonderfully creative and deeply humane. Consider the Orii Lab's "Dawn Cafe," a coffee shop staffed by small, remotely operated avatar robots. These robots are controlled by individuals with severe physical disabilities, allowing them to work, interact with customers, and earn an income from their homes or hospital beds. This isn't about replacing the barista; it's about augmenting human capability, allowing someone who was once isolated to rejoin the workforce and society in a meaningful way. It is the philosophy of augmentation brought to life.

The Demographic Imperative: Why Robots Are a National Necessity

Japan's embrace of robotics is not a hobby or a technological luxury. It is a matter of national survival. The country is navigating one of the most profound and rapid demographic shifts ever witnessed in the modern world. It is a "hyper-aging" society, where the number of elderly citizens is swelling while the birth rate continues to fall. This has created a perfect storm: a severe and ever-worsening labor shortage that threatens the very foundations of the nation's economy and social welfare systems.

This crisis has a name: the "2025 problem." This is the year when the last of Japan's post-war baby boomers—a massive generational cohort—will turn 75 or older. This single milestone is placing an almost unimaginable strain on the country's healthcare, pension, and long-term care systems. The numbers are stark and unforgiving. By 2040, it's estimated that nearly 35% of the population will be over 65. Looking further ahead, the national labor force is projected to plummet by as much as 24 million people by 2050. There simply will not be enough working-age people to fill all the jobs needed to keep the country running, from harvesting crops and building skyscrapers to stocking grocery shelves and caring for the elderly.

In this context, the conversation around AI and automation is completely different from the one taking place in many Western nations. In the West, the dominant fear is mass unemployment, where robots come to take jobs away from people. In Japan, automation is largely seen as a "potential savior." The fear is not an empty wallet; it's an empty factory, an unstaffed nursing home, an unmanned checkout counter. Technology is viewed as the most critical tool for sustaining economic productivity, maintaining essential public services, and ensuring a high quality of life for its cherished elderly citizens. This stark demographic reality is the engine driving massive investment, from both the government and the private sector, into the most advanced robotics and AI research on the planet.

The service sector, which is the backbone of the Japanese economy, accounting for about 75% of its GDP, offers a clear example. This sector is already facing an acute labor crisis. Retail stores, restaurants, and logistics companies are struggling to find enough workers to simply keep their doors open. This has created a powerful economic incentive to innovate. Companies are not adopting robotic solutions to fire workers and cut costs; they are doing it because they have no other choice. Convenience store chains like FamilyMart and Lawson are experimenting with robots that can stock drinks and ready-made meals on shelves, tasks that are physically demanding and hard to staff for overnight shifts. In warehouses, autonomous robots zip around, sorting and moving packages, ensuring that the logistics networks that power the nation's commerce don't grind to a halt.

This imperative is creating a new landscape of innovation across every part of Japanese life. In elder care facilities, you find robots like "Robear," a machine with a gentle, bear-like face designed to carefully lift residents from beds into wheelchairs, protecting caregivers from back injuries. There are smaller companion robots, like the fluffy, seal-like "PARO," which has been clinically shown to reduce stress and anxiety in elderly individuals, especially those with dementia. In the construction industry, where the workforce is also aging rapidly, companies are developing robotic systems to autonomously weld steel beams or pour concrete. In agriculture, where the average farmer is approaching 70 years old, automated tractors and drones that monitor crop health are becoming essential.

Ultimately, Japan's push into robotics is about something far more profound than economic figures. It's about maintaining a society's promise to its people. It's a pragmatic, deeply human effort to ensure that even with fewer people, the quality of life, the dignity of care, and the harmonious functioning of society can be preserved for generations to come. Japan is not just building robots; it is building its future.

The New Face of Care: How Japan is Humanizing its Robotic Revolution

In the quiet corridors of a Tokyo nursing home, the future has a gentle whir. It's not the cold, sterile future of science fiction, but something softer, more thoughtful. Here, the impact of Japan's national strategy is not an abstract concept but a daily reality. This is where AI and robotics are being woven into the delicate tapestry of elder care, a sector pushed to its limits by demographic change. The technologies being deployed are as diverse as the needs they serve, ranging from machines that provide a strong arm for lifting to companions that offer a soft touch for comfort. This isn't just about efficiency; it's a profound experiment in how technology can be used to amplify compassion, preserve dignity, and support both the caregivers and those they care for in the most vulnerable stages of life.

Robotics in Elder Care: From Companionship to Physical Assistance

The most visible and heartwarming applications are found in companionship and therapeutic robots. These are not meant to replace human interaction but to supplement it, filling the quiet hours and easing the pervasive loneliness that can settle in elder care facilities. The most iconic of these is PARO, a robotic baby harp seal. Covered in soft, anti-bacterial fur and filled with sensors, PARO responds to touch and the sound of a voice by cooing, wriggling, and blinking its soulful eyes. It learns to recognize a name and prefers to be petted in a way it remembers. Clinical studies have confirmed what caregivers have seen firsthand: interacting with PARO can significantly reduce stress, calm anxiety, and improve the moods of patients, particularly those grappling with the confusion of dementia.

Joining PARO are other social robots like PALRO, a small humanoid that acts as a cheerful activities director, capable of leading residents in morning exercise routines, hosting quiz games to keep minds sharp, and even holding simple conversations. Then there is LOVOT, a creature that looks like a cross between a penguin and an owl, designed with one purpose: to be loved. It roams around on wheels, seeks out people, and responds to affection with warm, happy gurgles. Its sole function is to create an emotional bond.

Beyond emotional support, a new class of robots provides crucial physical assistance. The work of a caregiver is physically punishing, and injuries from lifting and moving patients are a primary reason for staff burnout and turnover. Enter technologies like the HAL (Hybrid Assistive Limb). This powered exoskeleton, developed by the company Cyberdyne, is worn by a caregiver like a piece of high-tech clothing. It reads faint bio-electric signals on the wearer's skin, anticipates their movements, and provides a surge of robotic power to help them lift a resident with a fraction of the physical strain. Other versions of HAL are designed for the residents themselves, helping them to stand, walk, and participate in rehabilitation, giving them a renewed sense of mobility and independence. This category also includes clever innovations like robotic beds that fluidly transform into wheelchairs, eliminating the need for a risky transfer, and automated bathing systems that assist with hygiene while preserving a resident's dignity.

Overseeing it all are silent, AI-powered guardians. Advanced sensor systems are being deployed to enhance safety without the intrusive feeling of constant human surveillance. These systems can monitor a resident's movements and vital signs, using AI to instantly detect a fall or a sudden change in health and immediately alert the staff. This creates a safety net that is both constant and invisible, allowing for peace of mind while respecting personal space.

Vision-Language-Action (VLA) Models: The Next Wave of Intelligent Robots

A monumental leap forward in robotics is now underway, powered by what are known as Vision-Language-Action (VLA) models. This technology is what separates a simple factory robot from a truly intelligent helper. Think of it this way: a traditional robot is like a cook who can only follow a pre-written recipe, step-by-step, with no ability to deviate. A VLA-powered robot, on the other hand, is like a creative chef. You can walk into the kitchen and say, "Please make a healthy snack using the fruit on the counter," and the chef can see the fruit (vision), understand your request (language), and figure out the steps to wash, peel, and arrange it (action).

This ability to perceive, understand, and act in complex, ever-changing human environments is a game-changer. It is the key to unlocking the potential of robots in the dynamic settings where Japan needs them most: busy retail stores, crowded hospitals, and cluttered homes. For the first time, robots are being equipped with the common sense and adaptability that were once exclusively human traits. This technology directly confronts the labor shortages in Japan's critical service and logistics sectors by enabling robots to perform nuanced tasks that go far beyond simple, repetitive automation.

A powerful example of this technology in action is a retail robot prototype named Astra. Piloted in Japan to address severe staffing shortages, Astra is a glimpse into the future of service work. Imagine a store manager telling the robot, "The delivery just arrived; please restock the beverage aisle with the new products." A traditional robot would be stumped. But Astra springs into action. Its cameras (vision) scan the delivery boxes, identifying the new products. Its AI brain (language) processes the manager's command, understanding concepts like "beverage aisle" and "restock." Then, it formulates a plan (action).

Astra autonomously navigates through the store, deftly avoiding shoppers and displays. It locates the correct aisle and scans the shelves, identifying empty spaces. Its robotic arm then gently grasps a bottle and places it perfectly on the shelf. If it encounters an unexpected obstacle, like a fallen item, it can problem-solve, perhaps moving it aside before continuing its task. This fluid, intelligent response to a simple verbal command represents a quantum leap in automation, transforming a robot from a mere tool into a capable robotic colleague.

Cultural Acceptance and Thorny Ethical Questions

A crucial ingredient in Japan's robotic revolution is the country's unique cultural landscape, which has fostered a remarkable degree of public acceptance. This positive reception is often traced back to two key influences. The first is the prevalence of friendly, helpful robots in popular culture, from the iconic manga and anime hero Astro Boy to countless other characters who have portrayed robots not as threats, but as companions and protectors. Generations have grown up with this narrative, priming them to see robots in a positive light. The second is the subtle influence of Shinto religious beliefs, which include a form of animism that can ascribe a spirit or presence (kami) to inanimate objects. This worldview can make it feel more natural to form a bond with an object, whether it's a cherished tool or a sophisticated robot, blurring the hard line between the living and the mechanical.

This cultural predisposition is reflected in the data. One study found that an overwhelming 87% of elderly Japanese residents reported having positive experiences with care robots, and 92% felt safer knowing that automated monitoring systems were in place.

However, this warm embrace of technology is not absolute. It is accompanied by a thoughtful and ongoing ethical debate. The most persistent concern, voiced especially by veteran caregivers, is the potential for a "loss of warmth." They worry that an over-reliance on robotic efficiency could erode the vital human connection at the heart of caregiving. Can a robot, however sophisticated, truly replicate the empathy of a human touch, the nuance of a comforting word, or the shared understanding that comes from just sitting with someone in silence? One caregiver captured this central tension perfectly, expressing a cautious optimism about the helpfulness of robots but worrying that "the subtle communication, such as reading the room and understanding the situation, might be difficult to replicate." This captures the core dilemma: finding the right balance between the undeniable functional benefits of robotic assistance and the irreplaceable, relational quality of human care.

Japan's government is navigating this complex terrain with a characteristically thoughtful approach. Instead of imposing rigid, restrictive laws on AI development, it has promoted a "soft law" strategy. This involves creating non-binding guidelines and principles that emphasize human dignity, safety, privacy, and transparency. It fosters a collaborative environment where technologists, ethicists, government bodies, and the public work together to ensure that innovation aligns with deeply held societal values.

Practical Implications: A Blueprint for an Aging World

For policymakers across the developed world, from Europe to North America, who are watching their own populations begin to age, Japan offers a vital and proactive roadmap. It is a real-time case study in how to leverage AI and robotics not just to cope with a demographic shift, but to thoughtfully re-engineer a society to thrive within it. The primary lesson is the power of a comprehensive national strategy that combines significant government investment in research, fosters innovation in the private sector, and maintains a laser focus on solving real-world societal needs. Japan's success demonstrates that adopting these technologies is not a question of if, but of how.

For the technology developers and engineers who are building this future, the Japanese model highlights a crucial lesson: human-centered design is not a buzzword; it is the absolute key to success. In sensitive domains like healthcare and elder care, technical brilliance is not enough. The true challenge lies in creating technologies that are not just effective, but are also accepted, trusted, and integrated seamlessly and respectfully into the social fabric. This requires deep collaboration with the end-users—the elderly residents, the overburdened caregivers, and their families—to ensure that the robots being designed solve their problems in a way that feels helpful, not intrusive. The long-term success of a care robot depends less on its processing power and more on its ability to be a welcome and dignified presence in a person's life. Japan is showing the world that the most advanced technology is the kind that remembers the human at its center.

Estonia - AI and Democratic Processes: Combating Disinformation

Abstract

This chapter delves into the complex story of the Republic of Estonia, a nation celebrated globally as the "digital republic." It presents a fascinating, two-sided analysis of how artificial intelligence operates within one of the world's most hyper-digitized democracies. On one hand, Estonia stands as a beacon of progress. We will explore how the nation has ingeniously harnessed AI and built an incredible digital framework to elevate the very idea of citizenship. This isn't just about convenience; it's a model for transparent, efficient e-governance that actively encourages people to engage with their state. Imagine a country where tedious bureaucracy is nearly extinct, and public services are streamlined to the point of being almost invisible.

However, every light casts a shadow. On the other hand, this deep reliance on digital systems makes Estonia an incredibly attractive target for a new kind of warfare. The chapter examines the darker side of this coin, focusing on how Estonia's digital lifeblood exposes it to sophisticated, AI-powered disinformation campaigns and relentless cyber threats. These attacks aren't just technical glitches; they are malicious efforts designed to shake the foundations of its democratic processes. The core of this analysis will be dedicated to unpacking Estonia's remarkable and comprehensive strategy for survival and defense. It's a multi-layered, "whole-of-society" approach that builds a resilient shield against these modern threats. By studying Estonia, we can learn critical, real-world lessons about how a nation can protect its democratic soul in the turbulent age of artificial intelligence.

Theoretical Underpinnings

To truly grasp Estonia's unique situation, we need to look at it through two important theoretical lenses. The first is Democratic Theory, especially the ideas of a government that listens and citizens who participate. Think of Estonia's e-governance system not just as a collection of cool apps and websites, but as a deliberate technological experiment to bring old democratic ideals to life. Concepts like Deliberative Democracy, where citizens can easily discuss issues and influence policy, and Civic Engagement, where everyone has a voice, are baked into the system's DNA. The digital ID and online voting platforms are tools designed to lower barriers and encourage participation, making democracy more accessible and transparent for everyone.58

The second lens is Securitization Theory. This theory helps explain how a society can elevate a specific problem to the level of an existential threat, justifying extraordinary actions to fight it. For Estonia, this isn't just a theory; it's lived history. The massive, coordinated cyber-attacks in 2007 were a national trauma. This event, largely believed to have originated from Russia, was a brutal wake-up call.60 It forced Estonian leaders and citizens to "securitize" the digital world. Suddenly, cybersecurity and the threat of disinformation weren't just IT problems or media issues anymore. They were framed as fundamental threats to the nation's survival and independence. This act of securitization is the key to understanding Estonia's intense, all-encompassing response. It's the reason digital literacy is taught in schools like reading and writing, why the military has a robust cyber command, and why the government proactively hunts for election interference. Estonia's strategy isn't just a list of good ideas; it's a survival posture forged in the fire of a specific historical moment where protecting the digital realm became synonymous with protecting the nation itself.

E-Estonia: AI in a Digital-First Government

Since bravely stepping out from behind the Iron Curtain in the 1990s, Estonia has been on a relentless mission to build a government for the digital age. This ecosystem, known as e-Estonia, wasn't built overnight. It rests on a few brilliant, foundational pillars that have made it the perfect environment for artificial intelligence to flourish. At its heart is the X-Road, which you can think of as a secure digital highway system for data.64 It allows different government databases to talk to each other safely and instantly. Crucially, it means there is no single, giant government database that could be a tempting target for hackers; information remains distributed and protected.

Building on this is the mandatory Digital Identity (e-ID). This is far more than a simple login; it's a citizen's secure key to the entire country. Estonians use their e-ID card or Mobile-ID to do almost everything: file taxes in minutes, access personal health records, sign legally binding contracts, and even vote from their living rooms.58 This system is governed by the beautiful and simple "Once-Only" Principle.64 This law forbids the state from asking for the same piece of information twice. You provide your address once when you register, and every government agency that needs it can access it securely through X-Road. This simple rule eliminates endless paperwork and respects people's time.

This incredibly mature digital landscape is where AI now comes into play. Estonia's national AI strategy, nicknamed "#KrattAI," aims to make interacting with the government proactive and seamless.64 A centerpiece of this is the virtual assistant Bürokratt. Instead of trying to figure out which of the dozens of government agencies you need to contact, you can just ask your question in plain language. Bürokratt's network of AIs figures out who is responsible, finds the answer, and delivers it to you, aiming for a 24/7 government that works for its citizens, not the other way around.58 AI is also being used as a powerful tool for decision support. For example, the unemployment office uses AI to identify individuals at high risk of long-term joblessness and then suggests personalized training to help them get back on their feet, offering a helping hand before they fall too far.64

The Threat of AI-Driven Disinformation

Estonia's success story has a flip side. Its status as a proud digital pioneer and a frontline NATO member sharing a border with Russia makes it a permanent bullseye for foreign disinformation campaigns. The recent explosion in generative AI has thrown gasoline on this fire. This new technology allows adversaries to manufacture incredibly sophisticated and believable fake content on an industrial scale. We're talking about more than just fake news articles; we're talking about realistic deepfake videos, fabricated audio clips, and entire websites designed to look legitimate, all created in moments.59

These attacks have several clear and dangerous goals. First, they work to erode public trust. By constantly flooding social media and news feeds with conflicting and false narratives, they make citizens doubt their own government, the free press, and even each other. Second, these campaigns are designed to polarize society. They seek out sensitive social issues—cultural debates, historical grievances, economic anxieties—and pour fuel on them, hoping to amplify division and turn citizens against one another. Finally, and perhaps most directly, they aim to interfere with elections.61 Spreading lies about candidates or creating false stories about voting fraud can manipulate public opinion and damage the integrity of the democratic process.

Imagine this scenario: a week before a major election, a deepfake video surfaces. It appears to show a leading prime ministerial candidate making a racist and inflammatory remark. The video looks and sounds completely real. Within hours, it's shared hundreds of thousands of times across social media, amplified by an army of AI-powered bots. News organizations scramble to verify it, but by the time they can definitively prove it's a fake, millions of voters have already seen it, and the damage is done.68 This kind of AI-powered attack is a direct assault on the core of democracy, which relies on an informed public and a shared understanding of reality to function.

A Whole-of-Society Defense: Education, Fact-Checking, and Strategic Communication

Estonia learned a hard lesson, and they learned it early: you can build the strongest digital fortress in the world, with the highest firewalls and the most advanced encryption, but none of it matters if an adversary can trick your citizens into willingly opening the gates from the inside. This profound understanding is the very soul of their defense strategy. They recognized that the battle against disinformation isn't fought on servers, but in the minds of the people. This led to the creation of a "whole-of-society" defense, a philosophy that weaves resilience into the very fabric of the nation through three interconnected pillars.

The first and most important pillar is Education and Media Literacy. Estonia treats the ability to think critically about online information with the same seriousness as learning to read or do math. This isn't just a one-off class in high school; it is a concept woven into the national curriculum starting from a young age.60 Imagine a classroom where ten-year-olds are learning not just about historical events, but are also being taught how to spot a doctored photograph or question a viral video about that event. The goal is to create a kind of "cognitive inoculation." By teaching citizens how manipulation works—how emotional language is used, how sources can be faked, how algorithms create echo chambers—they are arming them with mental antibodies. This empowers Estonians to become skeptical consumers of information, habitually asking questions like, "Who is behind this message? Why are they showing it to me now? What information might they be leaving out?" before they ever click the "share" button.

The second pillar is a deep and genuine Collaboration with Civil Society. The Estonian government understands that it cannot, and should not, be the sole arbiter of truth. Instead, it actively supports and partners with a vibrant ecosystem of non-governmental organizations and volunteer groups who act as the nation's digital watchdogs.60 Groups like Propastop, staffed by passionate volunteers, have become expert "digital detectives," specifically tracking and exposing Russian propaganda campaigns in real-time.63 This isn't a top-down relationship; it's a partnership. The government can provide funding and high-level intelligence, while these civil society groups provide the agility, speed, and public credibility that official government channels sometimes lack. This model taps into Estonia's strong tradition of civic duty, turning national defense into a shared responsibility, not just a job for spies and soldiers.

Finally, the third pillar is swift and clear-eyed Strategic Government Communication. In the information age, silence is surrender. When a lie is racing around the world, the truth needs to get its boots on immediately. Estonian government agencies have abandoned slow, bureaucratic communication in favor of a proactive and rapid-response posture.61 During sensitive periods like elections, dedicated interagency task forces are spun up to monitor the information landscape 24/7. Their job is not only to debunk false stories as they appear but also to practice what is known as "pre-bunking"—anticipating likely lines of attack and proactively providing citizens with factual, verified information beforehand. By getting the truth out first and fast, they can fill the information space with credible content, leaving less room for disinformation to take root. Together, these three pillars create a powerful, living defense network that is far more resilient than any purely technological solution could ever be.

Ethical Governance of AI in Public Services

As Estonia builds its defenses against outside threats, it is simultaneously engaged in a profound and necessary internal conversation. With the immense power of artificial intelligence being woven into the government itself, a critical question emerges: How do you build a digital state that is not only efficient but also fundamentally good, fair, and just? This challenge—the ethical governance of AI—is central to maintaining the single most important asset Estonia possesses: the deep, hard-won trust of its people. The entire e-governance system only works because Estonians believe it works for them, transparently and without prejudice. Introducing AI, with its potential for hidden biases and opaque decision-making, puts this delicate trust equation to the ultimate test.

As a member of the European Union, Estonia's path is guided by robust legal frameworks like the GDPR and the forthcoming EU AI Act.66 These aren't just bureaucratic hurdles; they are essential guardrails designed to protect citizens. They enforce principles of data privacy, demand transparency for high-risk AI systems, and ensure that people have a right to understand and challenge automated decisions. However, rules alone are not enough. The real challenge lies in confronting the thorny problem of algorithmic bias. An AI system is only as good as the data it learns from. If an AI is trained on historical data from a time when societal biases were prevalent, it will learn and then perpetuate those same biases, often on a massive and unforgiving scale. An algorithm designed to predict who needs social assistance could, if not carefully designed and audited, end up discriminating against certain neighborhoods or minority groups, undermining the core democratic principle of equal treatment.66

Furthermore, there is the infamous "black box" problem. With some complex AI models, even their creators cannot fully explain how a specific input led to a specific output. This creates a terrifying accountability vacuum. If an AI denies someone a crucial public service, who is responsible? The programmer? The agency that used the tool? The data itself? To counter these risks, Estonia is focusing on establishing ironclad systems of human oversight, creating clear chains of accountability, and mandating regular audits of its algorithms to hunt for and root out bias.

This journey has involved learning from painful mistakes. A past scandal, in which a flawed automated system used for social benefits made biased decisions, served as a national wake-up call.66 It wasn't just a technical failure; it was a moral one that caused real harm and momentarily shook public confidence. This event, however, became a powerful catalyst for change. It forced a nationwide reckoning with the ethical stakes of AI and reinforced the absolute necessity of building systems that are transparent, accountable, and always, without exception, subject to meaningful human control. For Estonia, ensuring their own AI is ethical isn't just good policy; it is a national security imperative. If citizens cannot trust their own digital state, the entire edifice becomes vulnerable from within.

Practical Implications

The story of Estonia's digital journey offers a rich and vital blueprint for any democracy trying to find its footing in the turbulent 21st century. It's more than just an interesting case study; it's a living laboratory demonstrating how to defend democratic ideals against the insidious threat of AI-powered disinformation. The lessons learned are powerful, practical, and have profound implications for nations around the globe. The most important takeaway is perhaps the most surprising: building resilience is not primarily a technological challenge. While Estonia's digital infrastructure is world-class, its most effective defense is not found in code, but in the collective mind of its citizenry.

The clearest lesson is that the most durable shield against falsehood is a public that is difficult to fool. This requires a long-term, generational investment in public education focused on critical thinking and media literacy. Any country, regardless of its technological advancement, can begin this work tomorrow. It means redesigning curriculums to teach children not just what to think, but how to think—how to question sources, analyze motives, and understand the digital world they inhabit. This is the slow, patient work of building a human firewall, and it is the only defense that cannot be rendered obsolete by the next technological breakthrough.

A second critical implication is the power of building bridges between the government and civil society. In many Western nations, a degree of mistrust exists between state institutions and non-governmental actors. Estonia demonstrates that these groups are not adversaries but essential allies. A government can provide scale and resources, while civil society provides agility, niche expertise, and a level of public trust that state bodies may not always command. This collaborative model creates a national immune system where diverse and independent cells—from government task forces and volunteer fact-checkers to teachers and journalists—all work in concert to identify and neutralize informational threats before they can metastasize.

Finally, Estonia's experience teaches a vital lesson about leading by example. A nation cannot credibly fight for truth in the global information space if its own use of technology is opaque, unaccountable, or unfair to its own citizens. To build the public trust needed to withstand foreign propaganda, a government must first prove itself to be trustworthy. This means embracing radical transparency in its own use of AI, establishing clear ethical guidelines, and ensuring that automated systems serve the public, rather than control it. In the end, Estonia's message is both a warning and a source of hope. The fight against AI-driven disinformation is a permanent condition, not a war that can be definitively won. It requires constant vigilance, adaptation, and a deep commitment to the foundational principles of democracy. The most powerful lesson from the digital republic is that the resilience of a nation is ultimately a reflection of the critical engagement of its people.

Part III: Economic and Environmental Transformations

United Arab Emirates - AI and the Future of Work: Labor Displacement and Reskilling

The Inevitable Wave: Labor Displacement in Key Sectors

The United Arab Emirates' strategic embrace of artificial intelligence is not just about creating a new economic engine; it's about fundamentally re-engineering the existing one. This process, as Schumpeter's theory of creative destruction predicts, has a sharp edge. The "destruction" phase is already casting a long shadow over sectors that have long been the backbone of the nation's non-oil economy and major employers of its vast expatriate workforce. These are not distant, abstract threats but tangible shifts beginning to unfold.

Administrative and clerical roles are on the front lines. Tasks involving data entry, scheduling, document processing, and basic customer service are prime candidates for automation by sophisticated AI systems. For decades, these jobs have provided entry points for a diverse range of workers. Now, AI-powered chatbots, intelligent document processing software, and automated workflow systems are performing these functions faster, more accurately, and at a fraction of the cost. This represents an enormous efficiency gain for businesses but a profound challenge for the individuals whose livelihoods depend on these routines.

The logistics and transportation sector, a crown jewel of Dubai's economic model, is also facing a seismic shift. The vision of autonomous trucks navigating the E11 highway or drones delivering packages in dense urban areas is moving from science fiction to strategic planning. AI is optimizing supply chains, predicting maintenance needs for fleets, and managing port operations with minimal human intervention. While this boosts the UAE's position as a global logistics hub, it directly threatens the jobs of truck drivers, warehouse workers, and dispatch coordinators.

Similarly, the retail and customer service industries are being reimagined. AI-driven personalization engines are shaping what we buy online, while in-store experiences are increasingly mediated by technology. Cashier roles are being replaced by smart checkout systems. Customer support is heavily migrating towards AI chatbots that can handle a vast majority of inquiries, leaving only the most complex cases for human agents. Even roles in finance, particularly in areas like accounting, auditing, and financial analysis, are seeing routine tasks automated, pushing human workers to adapt or risk becoming obsolete. This isn't a simple replacement of one worker with one machine; it's a systemic overhaul where entire categories of work are being fundamentally redefined or erased.

Building the Future: The UAE's Ambitious Reskilling Agenda

Faced with this impending wave of displacement, the UAE is not standing idle. Its strategy is a direct and massive application of Human Capital Theory, treating its population's skills as the nation's most valuable asset in the AI era. The government is orchestrating a sweeping, top-down effort to reskill and upskill its workforce, aiming to transform citizens and residents from participants in the old economy to architects of the new one. This is not a scattered collection of training courses; it is a centrally planned, nationwide educational pivot.

A cornerstone of this effort is the "National Program for Coders," launched in partnership with global tech giants like Google, Microsoft, and Amazon. The program's ambitious goal is to attract and train 100,000 coders, establishing the UAE as a global hub for software development. It offers a golden visa residency track for top talent, provides access to cutting-edge training platforms, and fosters a community to support this new generation of digital builders. This initiative directly addresses the foundational skill needed to not just use AI, but to create and customize it.

Beyond coding, specialized institutions are leading the charge in advanced AI education. The Mohamed bin Zayed University of Artificial Intelligence (MBZUAI) in Abu Dhabi stands as a powerful symbol of this commitment. As the world's first graduate-level, research-based AI university, its mission is to cultivate a new generation of AI leaders, innovators, and thinkers right within the UAE. It attracts top students and faculty from around the globe, creating a powerful nexus of talent and research that seeds the local ecosystem.

Furthermore, the government is driving initiatives like "One Million Arab Coders," a vast online education platform that extends the reskilling mission beyond its borders, positioning the UAE as a leader in digital transformation for the entire Arab world. Private-public partnerships are also critical, with companies like G42 and Mubadala not only investing in technology but also establishing corporate academies and training programs to build a pipeline of talent equipped with practical, in-demand AI skills. The underlying philosophy is clear: the jobs of the future won't be found, they will be built, and building them requires a workforce armed with a completely new set of tools.

The Human Element: New Roles and Emerging Skill Demands

While the narrative often focuses on job losses, the "creation" side of Schumpeter's theory is equally dynamic. The integration of AI into the economy is not just eliminating roles; it is giving birth to entirely new professions and placing an unprecedented premium on a blend of technical and uniquely human skills. The UAE's job market is beginning to reflect this exciting and challenging transition.

Demand is soaring for technical specialists who can build, manage, and deploy AI systems. Roles like Machine Learning Engineer, Data Scientist, and AI Specialist are no longer niche positions but core components of modern business. These professionals are the ones designing the algorithms that power everything from personalized medicine to financial trading bots. Alongside them are AI Ops (or MLOps) engineers, who focus on the practical and complex task of maintaining and scaling AI models once they are in production, ensuring they run efficiently and reliably.

However, the new job landscape is not exclusively for deep-tech experts. A fascinating category of "hybrid" roles is emerging. These are positions that sit at the intersection of AI technology and traditional business functions. For example, an "AI Product Manager" needs to understand both the capabilities of machine learning and the needs of the market to guide the development of new AI-powered services. A "Prompt Engineer" specializes in the art and science of communicating with generative AI models to elicit the best possible outputs, a skill that is part creative, part technical.

Beyond the technical, there is a renewed emphasis on skills that machines cannot easily replicate. Critical thinking, complex problem-solving, and creativity are essential for navigating a world where basic information processing is automated. Emotional intelligence, communication, and collaboration become even more valuable as technology handles routine tasks, freeing up humans to focus on strategy, innovation, and managing complex stakeholder relationships. An "AI Ethics Officer," for instance, requires a deep understanding of not just technology but also philosophy, law, and social values to ensure that AI systems are deployed responsibly. The UAE's future workforce will need to be bilingual, fluent in both their professional domain and the language of technology.

Navigating the Labyrinth: Ethical Considerations and the New Social Contract

The UAE's bold, state-driven push into the AI future is efficient and decisive, but it raises profound ethical questions and fundamentally re-negotiates the social contract for both its citizens and its massive expatriate population. The speed and top-down nature of this transformation create unique challenges that more organic, market-led economies face more gradually.

A primary ethical concern revolves around the fate of the low- and semi-skilled expatriate workforce. These workers, who form the majority of the population, are heavily concentrated in the sectors most vulnerable to automation, such as construction, transportation, and retail services. The traditional social contract for this group has been transactional: labor in exchange for wages, with residency tied to employment. As AI displaces these jobs, what becomes of the people who hold them? The UAE's reskilling initiatives are largely focused on higher-skilled roles, creating a potential "lost middle" and raising questions about inclusivity and the state's responsibility to this vital segment of its population.

Data privacy and algorithmic bias represent another critical ethical frontier. The UAE's vision for AI-integrated "smart cities" and government services relies on the collection and analysis of vast amounts of citizen and resident data. This creates immense potential for efficiency but also significant risks. How is this data protected? Who has access to it? More importantly, are the AI models being trained on this data free from biases that could perpetuate or even amplify existing social inequalities? An algorithm designed to approve loans or screen job applicants could inadvertently discriminate against certain nationalities or demographics if not carefully designed and audited.

This centrally planned disruption forces a re-evaluation of the relationship between the state, the individual, and the economy. By framing the AI transition as a national strategic imperative, the disruption is presented as a necessary part of a collective mission. This powerful narrative can foster unity and purpose, but it also risks minimizing the real anxieties and hardships faced by those whose skills are devalued. Managing this transition ethically means creating robust social safety nets, ensuring equitable access to reskilling opportunities, and establishing strong governance frameworks to ensure that the pursuit of technological leadership does not come at the cost of human dignity and fairness.

Challenges on the Horizon: Potential Hurdles in the AI Race

While the UAE's ambition is backed by immense political will and financial capital, its path to global AI leadership is not without significant challenges and potential pitfalls. The sheer velocity of the strategy creates its own set of risks that could undermine its long-term success.

The most pressing challenge is the risk of a widening "skills chasm." The government can build universities and launch training programs, but reskilling an entire workforce takes time, and the pace of technological change is relentless. There is a real danger that the demand for high-end AI talent will outstrip the supply that can be cultivated locally, while the displacement of lower-skilled jobs happens much faster than those workers can be retrained for new roles. This could lead to an increased reliance on importing foreign experts to fill senior positions, while a segment of the existing population is left behind, exacerbating social and economic inequality.

Another significant hurdle is the dependency on foreign technology. While the UAE is investing heavily, the core intellectual property for foundational AI models and advanced semiconductors (like the Nvidia chips mentioned in the US-UAE partnership) still resides primarily in the United States and China. The UAE's strategy of deep partnership is a clever way to secure access to this technology, but it also creates geopolitical vulnerabilities. Shifts in global trade policy, tech export controls, or international relations could potentially disrupt the UAE's access to the critical hardware and software that underpins its entire AI vision.

Finally, there is the challenge of fostering a genuine, bottom-up innovation culture. A top-down, state-led model is excellent for building infrastructure and directing capital, but true, disruptive innovation often bubbles up from a more chaotic ecosystem of startups, independent researchers, and risk-taking entrepreneurs. Can a centrally planned strategy create the dynamic, flexible, and sometimes messy environment where groundbreaking ideas flourish? The UAE must balance its grand national projects with policies that empower individual creators and small businesses to experiment, fail, and innovate. Successfully navigating these challenges will determine whether the UAE's grand AI experiment becomes a global model for economic transformation or a cautionary tale of a future that arrived too fast.

The Human Face of Displacement: A Deeper Look at At-Risk Sectors

The statistics on AI-driven job displacement, like Goldman Sachs's estimate of 300 million affected jobs, can feel abstract and distant. But within the gleaming towers and bustling souks of the United Arab Emirates, this transformation has a distinctly human face. The sectors identified as "at-risk" are not just lines in an economic report; they are the livelihoods of millions of people, particularly the nation's vast and vital expatriate workforce.

Let's look closer at administrative roles. For decades, the smooth operation of the UAE's booming businesses has relied on an army of office administrators, receptionists, and executive assistants. These positions have been a crucial entry point for skilled workers from South Asia, the Philippines, and other parts of the world. Now, AI-powered productivity suites can manage calendars, draft emails, transcribe meetings, and organize files with startling efficiency. An AI can do the work of several people, and it never takes a sick day. This isn't just about making offices more efficient; it's about a fundamental shift that could hollow out a significant portion of the white-collar service economy.

Consider the vibrant retail and customer service landscape. From the sprawling malls of Dubai to the call centers in Sharjah, providing excellent service has been a hallmark of the UAE's brand. Yet, AI is rapidly changing this. Sophisticated chatbots, powered by natural language processing models, now handle customer queries with patience and accuracy, 24/7. In-store, automated checkout systems and inventory management bots reduce the need for cashiers and stock clerks. This directly impacts a massive workforce, often comprising individuals for whom these jobs are a critical lifeline to support families back home.

Even in the logistics hubs that are the pride of the nation, change is in the air. At Jebel Ali Port, AI algorithms optimize the movement of containers, predicting the most efficient routes for cranes and trucks. In warehouses across the country, autonomous mobile robots now zip through aisles, picking and packing orders with a speed that humans simply cannot match. While this technological prowess strengthens the UAE's position as a global trade leader, it simultaneously diminishes the value of the manual labor that built this industry, creating a profound sense of uncertainty for the drivers, pickers, and sorters who keep the goods moving.

Forging a Future-Ready Workforce: A Multifaceted Education Strategy

The UAE's leadership understands that resisting this technological tide is futile. Instead, they have chosen to build a new kind of ark. Their response to the challenge of displacement is a massive, multi-layered investment in human capital, designed to equip their people with the skills to navigate the coming flood. This is far more than just a single university; it is a national reorientation of education.

The Mohamed bin Zayed University of Artificial Intelligence (MBZUAI) is undoubtedly the glittering flagship of this strategy. It is a bold statement, designed to attract the world's brightest minds and create a gravitational center for AI research and development within the region. Its partnerships, like the one with IBM, are not just academic exercises. They are strategically aimed at solving real-world problems critical to the UAE's future, such as water scarcity and energy efficiency, while simultaneously building a homegrown talent pool fluent in the most advanced AI techniques.

But the strategy goes much deeper. Recognizing that the future needs builders, not just thinkers, the government launched the "National Program for Coders." This is a nationwide mobilization effort, a digital call to arms, with the audacious goal of training 100,000 coders. It partners with global tech giants like Google and Microsoft to offer accessible training, creates pathways to employment, and even offers long-term "Golden Visas" to attract top programming talent from around the world. It's a pragmatic and powerful initiative aimed at building the digital infrastructure of tomorrow.

This future-proofing extends all the way down to K-12 education. Curriculums are being revamped to include computational thinking, robotics, and data literacy from an early age. The goal is to cultivate a generation of "digital natives" who are not just users of technology, but creators and innovators. Initiatives run by entities like the Dubai Future Foundation further supplement this by running workshops, challenges, and accelerator programs that expose young people and existing professionals to frontier technologies, fostering a culture of lifelong learning and adaptation. This is not just about preparing people for jobs; it's about cultivating an entire society that is resilient, adaptable, and ready to thrive in an age of intelligent machines.

The Dawn of New Professions: Job Creation in the AI Economy

While the narrative of automation is often tinged with fear, the other side of this coin is one of immense opportunity. The AI revolution is not just an engine of destruction; it is a powerful force of creation, giving rise to entirely new job categories and career paths that were unimaginable just a decade ago. The UAE is actively positioning itself to become a hub where these new professions take root and flourish.

The most obvious new roles are deeply technical. The demand for Data Scientists, who can interpret vast datasets to find actionable insights, and Machine Learning Engineers, who can design and build the intelligent algorithms themselves, is exploding. But other, more specialized roles are emerging too. AI Ops Engineers are needed to ensure that these complex AI systems run smoothly and reliably at scale. AI Security Specialists are tasked with protecting these systems from new forms of cyberattacks that target algorithmic vulnerabilities.

Beyond these core technical roles, a fascinating ecosystem of "hybrid" jobs is blooming. These positions bridge the gap between the technology and its real-world application. For example, an AI Ethicist works with companies to ensure their AI systems are fair, transparent, and aligned with human values. This requires a unique blend of technical understanding, legal knowledge, and philosophical insight. A Prompt Engineer specializes in the craft of communicating with generative AIs like ChatGPT, figuring out the precise questions and commands that yield the most useful and creative results. This is less about coding and more about language, logic, and creativity.

Crucially, this new economy places a huge premium on skills that are profoundly human. As AI takes over routine analytical and administrative tasks, the value of skills like critical thinking, complex problem-solving, emotional intelligence, and collaboration skyrockets. The jobs of the future will require people who can ask the right questions, not just find the answers; who can lead teams of diverse individuals to solve ambiguous problems; and who can communicate complex ideas with clarity and empathy. The successful professional in the UAE's AI-driven future will be the one who can work with intelligent systems, leveraging their power to augment their own innate human creativity and ingenuity.

Rethinking the Social Fabric: AI's Impact on Society and Inequality

The UAE's rapid, top-down AI implementation is an impressive feat of national will, but it also places immense pressure on the country's unique social fabric. This technological restructuring is not happening in a vacuum; it is happening in a nation where expatriates make up over 85% of the population, and where the social contract has long been based on a clear exchange of labor for residency. AI threatens to upend this long-standing arrangement, raising critical questions about equity and the future of society.

The most significant challenge is the growing risk of a digital divide turning into a societal chasm. The government's excellent reskilling programs are primarily aimed at creating a high-skilled knowledge economy. But what about the bus driver, the retail worker, or the construction laborer whose job is automated away? For many of them, a graduate-level course in machine learning is not a realistic pathway. Without clear and accessible programs for "transitional" upskilling—training for roles like robot maintenance technicians or smart building supervisors—there is a real danger of creating a two-tiered society: a high-paid, AI-literate elite and a displaced class of workers with obsolete skills.

This raises fundamental questions about the social safety net. In a system where residency is tied to employment, what happens when large numbers of jobs are permanently eliminated? This is a profound ethical challenge for the government. It requires a rethinking of social responsibility, potentially exploring new models for unemployment support, portable benefits, or pathways to more permanent residency for long-term contributors to the economy who are displaced by technology.

Furthermore, the concentration of AI development in the hands of a few large, state-backed entities like G42 creates a risk of economic power becoming even more centralized. Ensuring that the incredible wealth and productivity gains generated by AI are distributed broadly across society—and not just captured by a small group of investors and technologists—is essential for long-term social stability. The UAE's journey into the AI future is as much a social and ethical experiment as it is a technological one. Successfully navigating it will require not just smart policy, but a deep commitment to inclusivity and human dignity.

Beyond Buzzwords: Unpacking the UAE's AI Ethics in Practice

The UAE's commitment to an ethical AI transition, as outlined in its charters and policies, is a crucial acknowledgment that technological progress cannot be divorced from human values. The principles of fairness, transparency, and accountability are easy to state but incredibly complex to implement, especially in a society as uniquely diverse as the UAE. Unpacking what these principles mean in practice reveals the profound challenges and ambitions of the nation's strategy.

What does "fairness" truly mean in an algorithm used for hiring in a country with over 200 nationalities? An AI trained on historical hiring data might inadvertently learn and amplify existing biases, perhaps favoring candidates from certain countries or educational backgrounds. Ensuring fairness requires a proactive and continuous effort to audit these systems, test them for discriminatory outcomes, and use techniques to de-bias the data they learn from. For the UAE, this isn't just a technical problem; it's a matter of upholding the multicultural harmony that is central to its national identity. A biased algorithm doesn't just make a bad decision; it threatens the very fabric of social cohesion.

"Transparency" and "explainability" present another frontier. When an AI system denies someone a loan, flags them for a secondary security screening, or recommends a particular medical treatment, a simple "computer says no" is not enough. The principle of explainability demands that the system's decision-making process can be understood by a human. This is vital for building trust and providing a mechanism for appeal. For the government and businesses in the UAE, this means investing in "Explainable AI" (XAI) technologies and training staff who can interpret and communicate these complex decisions to the public. It's about shifting from a "black box" approach, where the AI's logic is opaque, to a "glass box" approach that fosters accountability.

Ultimately, the principle of being "human-centered" serves as the guiding star. It insists that AI should augment human capabilities, not simply replace them. It means designing systems that empower doctors to make better diagnoses, help teachers personalize learning for students, and free up public servants from monotonous paperwork to focus on serving citizens more effectively. It's a commitment to ensuring that this grand technological project serves the well-being of the people, keeping human dignity and autonomy at the heart of the automated future.

Winning Hearts and Minds: The Campaign for Public Trust

The UAE's AI strategy is not just a technological or economic plan; it is a societal one. Its success hinges not only on developing sophisticated algorithms and reskilling workers but also on securing the most crucial element of all: public trust. The government understands that a population that is fearful or suspicious of AI will resist its adoption, undermining the entire national vision. Consequently, a quiet but deliberate campaign to win the hearts and minds of the public is underway.

This effort goes far beyond publishing policy documents. It involves a concerted push towards demystification. Government bodies and affiliated organizations like the Dubai Future Foundation regularly host exhibitions, workshops, and public forums designed to make AI tangible and understandable. They showcase practical applications—from AI-powered traffic management to personalized healthcare services—that directly benefit residents' daily lives. This strategy is simple yet effective: it transforms AI from an abstract, job-destroying threat into a helpful tool that makes life easier, safer, and more efficient.

Transparency is a key pillar of this trust-building exercise. The government's proactive publication of its ethical charters and data protection laws is a signal to the public that there are rules to this new game. It's a promise that the development of AI will not be a lawless digital wild west. By establishing clear principles for accountability and data privacy, the state is attempting to build a foundation of psychological safety, assuring residents that their information will be protected and that the technology will be held to a high standard.

Furthermore, the narrative surrounding AI is carefully curated. It is consistently framed not as a story of replacement, but as one of empowerment and national progress. The focus is on the new opportunities being created, the global leadership being asserted, and the role of AI in solving major challenges like climate change and disease. By inviting the public to see themselves as participants in this exciting national project—the journey to a post-oil future—the government aims to foster a sense of shared purpose, turning potential anxiety into collective ambition.

Weaving a New Safety Net: Redefining the Social Contract

Perhaps the most critical and challenging aspect of the UAE's AI transition is the need to re-imagine its social contract. The existing model, particularly for the vast expatriate population, has been straightforward: your right to live and work in the country is tied directly to your job. As AI begins to erode the stability of entire job categories, this model becomes fragile. The practical implication is that the UAE must now weave a new kind of social safety net, one designed for the realities of a more fluid and uncertain labor market.

This goes far beyond traditional unemployment benefits, which have not been a central feature of the expatriate labor system. The conversation is shifting towards new forms of support. One potential innovation is the concept of "transitional income support," providing a temporary financial cushion for workers who are displaced by automation while they undergo government-approved reskilling programs. This would be a radical shift, but a necessary one to prevent widespread hardship and ensure that displaced workers have the breathing room to successfully retool for new careers.

Another crucial thread in this new safety net is the idea of portable, internationally recognized skills certifications. As workers move between jobs—and potentially between countries—their skills and qualifications need to be fluid. The government can play a vital role in establishing blockchain-based digital credentials or "skill wallets" that allow a worker to carry a verified record of their training and competencies with them. This would empower individuals to better market themselves in a dynamic job market and ensure their investment in reskilling is not lost if they change employers.

Ultimately, this transition forces a deeper conversation about the role and responsibility of the state towards its entire resident population, not just its citizens. Managing the disruption caused by AI in a just manner requires a new social bargain. It means building systems that provide not just opportunity, but also security. It's about ensuring that the incredible prosperity unlocked by artificial intelligence doesn't leave behind the very people whose labor helped build the country.

A Model for the World? The UAE's Blueprint and Its Unique Context

The UAE's proactive, state-led approach to the AI revolution is undeniably a compelling case study, offering valuable lessons for nations across the globe. The core takeaway is powerful in its simplicity: you cannot embrace the opportunities of AI without simultaneously addressing its social consequences. The UAE's model demonstrates the necessity of pairing technological ambition with a deep and well-funded commitment to human capital development.

Other countries can learn from the UAE's strategic foresight. By anticipating disruption instead of reacting to it, the nation has bought itself valuable time to build the educational institutions and reskilling programs needed for the transition. The close synergy between government vision, sovereign wealth fund investment, and public-private partnerships creates a powerful, unified engine for change that is difficult to achieve in more fragmented economies. This integrated approach, where industrial policy, education policy, and technology policy are all pulling in the same direction, is a model of effective governance in the 21st century.

However, it is crucial to recognize the caveats. The UAE's model is not a one-size-fits-all blueprint that can be easily replicated. Its success is deeply intertwined with its unique political and economic structure. As a federation of monarchies with a highly centralized decision-making process, the government can enact sweeping, long-term strategies without the political friction common in many democracies. Furthermore, its ability to fund this transition is fueled by immense sovereign wealth from decades of hydrocarbon exports. Few nations have the political agility or the deep financial reserves to direct a hundred-billion-dollar fund toward a single technological goal.

Therefore, while other nations should be inspired by the UAE's ambition and its holistic approach, they must adapt the principles to their own realities. For a European nation, it might mean leveraging the EU's regulatory power to guide ethical AI development. For a developing nation, it could involve focusing on mobile-first educational platforms to deliver skills at scale. The lesson from the UAE is not to copy its exact methods, but to embrace its core philosophy: that managing the future of work requires deliberate, strategic, and profoundly human-centered planning.

South Korea - AI in Financial Services: Fairness in Lending and Credit

Introduction: The Digital Pulse of a Financial Powerhouse

South Korea is a nation that lives and breathes technology. From the hyper-connectivity of its cities to the global dominance of its tech giants, the country operates at a pace few can match. It's no surprise, then, that its financial sector has embraced artificial intelligence with breathtaking speed. In the bustling financial districts of Seoul, AI is not a futuristic concept; it is the engine of the modern economy. It works silently in the background, flagging fraudulent transactions, personalizing investment advice, and, most critically, deciding who gets a loan.

This rapid adoption of AI has unlocked incredible efficiency. Banks and FinTech startups can now process loan applications in minutes, not days, using algorithms to assess risk with a precision that was once unimaginable. This technological leap promises a more inclusive financial system, one where more people can access the credit they need to buy a home, start a business, or manage an emergency.

However, this shiny new world of algorithmic finance casts a long shadow. What happens when the algorithm says no? And more importantly, why did it say no? As South Korea races forward, it is confronting a profound ethical dilemma. The very AI systems designed to be objective and efficient carry the risk of becoming digital gatekeepers, capable of creating new forms of discrimination and deepening old social divides. This chapter explores South Korea's journey as it navigates this complex terrain, attempting to build a future where financial innovation serves everyone fairly.

The Peril of the Black Box: Algorithmic Bias in Lending

At its core, an AI lending model is a powerful pattern-recognition machine. It sifts through vast amounts of data—far more than any human ever could—to predict who is likely to repay a loan. This data includes traditional financial information, like income and debt levels. But increasingly, it also includes alternative data, such as online shopping habits, bill payment history, and even social media activity. The promise is that by looking at a wider picture, the AI can find creditworthy individuals who might have been overlooked by old-fashioned credit score systems.

The danger lies in what these patterns might represent. AI learns from historical data, and if that data reflects existing societal biases, the AI will learn those biases, too. For example, if a certain low-income neighborhood has historically had a higher loan default rate, the algorithm might learn to penalize any applicant from that area, regardless of their individual financial stability. This creates a vicious cycle known as "digital redlining," where technology effectively draws discriminatory lines around communities, locking them out of economic opportunities.

The problem is made worse by the "black box" nature of many advanced AI models. Often, even the developers who build the algorithms cannot fully explain the precise logic behind a specific decision. An AI might deny a loan based on thousands of subtle correlations in the data, making it impossible to point to a single, clear reason. This lack of transparency is a massive problem for both consumers and regulators. A person denied credit has a right to know why, and if no one can provide a clear answer, there is no way to challenge a potentially unfair or biased decision.

Ethical Foundations: Distributive Justice and Equal Opportunity

The challenge of algorithmic bias is not just a technical problem; it is a profound ethical one. It forces us to ask fundamental questions about what a fair society looks like. These questions are at the heart of philosophical concepts like Theories of Distributive Justice, which are all about the fair allocation of a society's resources and opportunities. Access to credit is one of the most important resources in a modern economy. It is the key to social mobility. If AI systems distribute this resource in a way that unfairly benefits some groups and harms others, they become instruments of injustice.

This connects directly to the principle of Equal Opportunity. This idea holds that people should be judged on their own merits and efforts, not on characteristics they cannot control, such as their family background, where they live, or their age. When an AI lending model uses data that acts as a proxy for these characteristics, it violates this principle. For instance, an algorithm might not explicitly use race or gender as an input, but it might learn that people who shop at certain stores or have certain hobbies are higher risks. If these behaviors correlate strongly with a particular demographic group, the result is the same: a system that discriminates.

South Korea's efforts to regulate AI in finance can be understood as an attempt to embed these ethical principles into its laws. The goal is to ensure that the move to algorithmic decision-making doesn't leave behind the country's commitment to fairness. It is an effort to translate the timeless value of equal opportunity into the language of code, ensuring that the future of finance is built on a foundation of justice, not just efficiency.

The "Dual-Track" Strategy: Fueling Innovation, Building Trust

Confronted with this challenge, the South Korean government has chosen a path that is both ambitious and pragmatic. It has adopted what can be described as a "dual-track" strategy, which aims to simultaneously promote rapid innovation while carefully putting regulatory guardrails in place. It's a calculated effort to have the best of both worlds: a globally competitive FinTech industry and a financial system that consumers can trust.

On the first track, the government is actively pushing for AI adoption. It provides funding, supports research, and creates "regulatory sandboxes." These sandboxes are special programs where FinTech companies can test new and innovative AI-powered services on real consumers in a controlled environment, temporarily free from certain regulations. This pro-innovation stance is driven by a clear economic imperative. South Korea sees AI as essential for its future growth and is determined to maintain its position as a global technology leader.

On the second, parallel track, the government is building a framework for accountability. Policymakers understand that technology can't succeed in a vacuum. Long-term public adoption of AI-driven finance depends entirely on trust. People will not use systems they believe are unfair, secretive, or biased against them. Therefore, this second track is focused on creating rules that ensure transparency, fairness, and consumer protection. This dual approach reveals a core belief in South Korea's governance model: that responsible regulation is not an obstacle to innovation, but a necessary ingredient for its sustainable success.

The Rulebook for Robots: The AI Basic Act and FSC Guidelines

The centerpiece of South Korea's regulatory effort is the landmark AI Basic Act. This law is not designed to regulate every single use of AI. Instead, it takes a risk-based approach, focusing its most stringent requirements on "high-impact" AI systems. These are defined as systems that could significantly affect public safety and fundamental human rights. Unsurprisingly, AI used for credit scoring and loan applications falls squarely into this category.

For these high-impact systems, the AI Basic Act imposes several key obligations. Companies using them must ensure their systems are transparent, meaning they need to be able to provide a basic explanation for their decisions. They must conduct fairness assessments to check for and mitigate biases. And they must establish clear lines of accountability, so there is no question about who is responsible when something goes wrong. However, in a nod to its innovation-first approach, the penalties outlined in the Act are more modest compared to those in Europe's comprehensive AI Act. This signals a desire to guide and encourage compliance rather than to punish, reflecting a strategy to regulate without stifling the experimentation vital to a growing tech ecosystem.

To complement this broad legal framework, the Financial Services Commission (FSC) has issued more detailed guidelines specifically for the financial sector. These guidelines get into the practical details, advising banks and FinTech firms on how to manage their data, how to design and validate their AI models to check for bias, and how to create clear and understandable explanations for consumers who are denied credit. The FSC's work is crucial because it translates the high-level principles of the AI

Basic Act into concrete, actionable steps for financial institutions.

Conclusion: A Delicate Balance for a Digital Future

South Korea stands at a critical juncture. It is enthusiastically embracing the power of artificial intelligence to create a more efficient and dynamic financial system. At the same time, it is acutely aware of the risks that this powerful technology poses to fairness and social equity. The country is engaged in a delicate and ongoing balancing act—a high-stakes effort to nurture its vibrant FinTech sector while weaving a strong safety net of consumer protection and ethical oversight.

The "dual-track" strategy, anchored by the AI Basic Act and the detailed guidance of the FSC, represents a uniquely Korean approach to a global problem. It is pragmatic, forward-looking, and built on the understanding that innovation and trust must go hand in hand. The journey is far from over. As AI models become more complex and the use of alternative data grows, regulators and companies will face new and unforeseen challenges. The world is watching South Korea, not just as a leader in technology, but as a test case in the quest to build a digital financial future that is not only smarter and faster, but also fundamentally fair. The lessons learned here will echo far beyond its borders, offering a potential blueprint for other nations grappling with the same algorithmic dilemma.

Part I: The Rise of the AI-Powered FinTech Revolution

South Korea's financial industry has always been fast, but the arrival of artificial intelligence has pushed it into hyperdrive. The old world of banking—defined by paperwork, long queues, and slow, deliberate decisions—

is rapidly fading in the rearview mirror. In its place, a new ecosystem is emerging, one that is smarter, faster, and deeply interwoven with the fabric of digital life. This isn't just a minor upgrade; it's a full-scale revolution, and AI is the engine powering it all. Financial institutions, from towering legacy banks to nimble FinTech startups, are deploying AI in ways that are fundamentally changing how people save, borrow, and invest.

At the very heart of this transformation is algorithmic credit scoring. For generations, your ability to get a loan was determined by a fairly rigid set of criteria: your salary, your existing debts, and your history with a few major credit bureaus. This system worked, but it had massive blind spots. It often failed to see the full picture, leaving millions of "thin-file" customers—like young people, freelancers, or recent immigrants with little credit history—out in the cold. AI has shattered this old model. Today's algorithms look at a universe of data points. They don't just see your salary; they see that you've paid your phone and utility bills on time for five straight years. They might analyze your online purchasing behavior or other digital footprints to build a far more nuanced and holistic portrait of your reliability. The promise is profound: to unlock financial opportunities for millions who were previously invisible to the system, judging them on the reality of their lives, not the limitations of old data.

While AI is opening new doors for customers, it's also building stronger walls against criminals. Fraud detection has become one of the most powerful applications of machine learning in finance. Old systems were based on simple, rigid rules. For example, a rule might automatically flag any transaction over a certain dollar amount or one made from a foreign country. These systems were easy for sophisticated fraudsters to outsmart. An AI, however, works like a digital detective. It learns the unique rhythm of your financial life—where you shop, what time you buy groceries, how much you typically spend. It recognizes patterns that a human would never notice. When a transaction suddenly breaks that pattern, the AI flags it instantly. It can tell the difference between you buying a plane ticket for a vacation and a thief using your card details on another continent, stopping crime in its tracks far more effectively than any human team ever could.

This new intelligence is also making finance more personal and accessible. AI has given rise to a new generation of personalized financial services. Think of "robo-advisors," which are essentially AI-powered investment managers. In the past, getting professional investment advice was a luxury reserved for the wealthy. Now, anyone with a smartphone can access an algorithm that will analyze their financial goals and risk tolerance to build a diversified investment portfolio for them, automatically managing it over time. This isn't just about investing; AI is personalizing everything, from recommending the credit card with the best rewards for your spending habits to offering tailored savings plans. It's making financial management less intimidating and empowering a whole new generation to take control of their economic futures.

The explosive growth of these AI-driven services would not have been possible without one critical development: open banking. In 2019, South Korea launched a standardized "open banking API platform." An API, or Application Programming Interface, is like a secure, universal translator that allows different software programs to talk to each other. This platform essentially forced all the major banks to create a secure doorway through which customers could allow third-party apps to access their financial data.

The change was seismic. It was a fundamental rewiring of the relationship between banks, developers, and customers. The results speak for themselves. By the end of 2022, the platform had seen an astonishing 160 million new subscribers and 350 million registered accounts. This unleashed a tidal wave of innovation. FinTech companies could suddenly build apps that let users see all of their bank accounts in one place, transfer money between different banks with a single tap, or use AI to analyze their complete financial picture to offer smarter budgeting advice. Open banking became the fertile ground where South Korea's AI-powered FinTech scene could truly blossom.

Part II: The Specter of Algorithmic Bias in Finance

For all its incredible promise, the AI revolution in finance has a dark side. There is a ghost in the machine. While these algorithms are designed to be objective and impartial, they have the potential to become powerful engines of discrimination, creating new forms of inequality that are harder to see and even harder to fight. The very data that fuels their intelligence can also be their poison.

This risk is most acute in lending and credit. An algorithm is not born with an understanding of fairness; it only knows what it is taught from the data it is fed. If that training data is a reflection of a society with a long history of bias, the AI will learn, replicate, and even amplify those same prejudices. This can lead to a deeply troubling phenomenon known as "digital redlining." The term "redlining" comes from the historical practice where banks would draw red lines on a map around minority neighborhoods and systematically deny them loans. Digital redlining is the modern equivalent. An algorithm, trained on biased historical data, might learn that applicants from a certain neighborhood, income bracket, or even ethnic background are "higher risk." It doesn't draw a red line on a map; it builds an invisible fence out of code, systematically denying opportunities to entire groups of people, regardless of their individual merit.

To understand how this happens, it's crucial to recognize the different ways bias can seep into an algorithm. The most common is pre-existing bias. This is bias that already exists in society and is captured in the data. Imagine an AI model trained on decades of loan application data from a time when societal prejudices were even more pronounced. The data might show, for example, that women were approved for smaller business loans than men, or that people from a specific racial group were denied mortgages at a higher rate. The AI, in its effort to find patterns, will learn these discriminatory outcomes as a rule. It will conclude that being a woman or belonging to that racial group is a risk factor, and it will carry that historical injustice forward into the future.

Then there is technical bias. This is a bias that is accidentally introduced during the design and development of the AI. It's less about societal prejudice and more about flawed methodology. For example, if developers train a credit-scoring model using data collected primarily from wealthy, urban areas, the model will become an expert on that specific demographic. When it later has to evaluate an applicant from a rural or lower-income area, it may struggle to understand their financial profile and unfairly judge them as a higher risk. The model isn't intentionally biased, but because its "education" was incomplete and unrepresentative, its decisions become skewed.

Finally, there is the subtle but powerful danger of emergent bias. This is bias that develops over time as the AI system operates in the real world and creates feedback loops. Think of a product recommendation engine on a shopping site. It learns to promote what is popular. As it promotes those popular items, they become even more popular, and the AI promotes them even more. Over time, less popular or niche products become almost invisible. A similar feedback loop can occur in finance. An algorithm might start with a slight bias towards applicants with a certain profile. As it approves more of these "safe" applicants and they successfully repay their loans, the AI becomes even more confident that this profile is the ideal. It becomes a self-fulfilling prophecy, gradually narrowing its definition of "creditworthy" and marginalizing anyone who doesn't fit into its increasingly rigid box.

The real-world consequences of this are devastating. Consider a loan approval algorithm trained on biased historical data. A highly qualified entrepreneur from a historically disadvantaged group applies for a startup loan. She has an excellent business plan, a strong income, and a perfect repayment history. But the algorithm, sifting through its data, picks up on subtle correlations linked to her background—correlations that it has learned to associate with risk. It denies the loan. For the customer, this is a crushing, often inexplicable rejection that kills a dream. For the financial institution, it is a catastrophic failure. It leads to customer anger, immense reputational damage, and ultimately, regulatory backlash. It is the moment when the promise of AI—to create a more objective and inclusive financial system—completely and utterly breaks down.

Part I: South Korea's Regulatory Response: The AI Basic Act and FSC Guidelines

Faced with the immense promise and potential peril of artificial intelligence, South Korea did not stand still. Instead of letting the technological revolution unfold without rules, the government and regulators have moved to become its architects, designing a blueprint for responsible AI. They are building a framework intended to channel the power of AI for economic growth while constructing strong guardrails to protect citizens from its risks. This response is not a single, simple rule but a sophisticated, multi-layered strategy designed for a complex new era.

The cornerstone of this entire structure is the landmark Act on the Development of Artificial Intelligence and Establishment of Trust, more simply known as the "AI Basic Act." Passed by the National Assembly in the final weeks of 2024 and set to take effect in January 2026, this is South Korea's foundational law for the age of algorithms. Rather than applying a blunt, one-size-fits-all approach, the Act embraces a modern, risk-based philosophy. It understands that an AI used to recommend music requires far less scrutiny than one that decides whether you can buy a home.

To put this philosophy into practice, the AI Basic Act categorizes AI systems based on their potential to impact people's lives. At the top of this pyramid are the "high-impact" AI systems. These are the powerful algorithms that make critical decisions affecting fundamental rights and opportunities. The law is crystal clear: financial systems used for credit scoring and loan screening fall directly into this high-stakes category. With great power comes great responsibility, and the Act spells out precisely what that responsibility entails.

Operators of these high-impact systems are no longer free to hide behind a "black box." They are now bound by a specific set of obligations. They must conduct thorough impact assessments before deploying a new AI, proactively identifying and planning to mitigate potential risks of bias or harm. They are required to maintain detailed records of the measures they take to ensure the safety and reliability of their models. Crucially, they face a new mandate for transparency, meaning they must be able to disclose the main criteria an algorithm uses to make its decisions. And perhaps most importantly, they must provide clear routes for redress and human review, ensuring that a person is always in the loop when it matters most.

While the AI Basic Act provides the overarching legal skeleton, the Financial Services Commission (FSC) has added the muscle and sinew. The FSC has issued sector-specific guidelines that translate the high-level principles of the Act into a practical playbook for banks, insurance companies, and FinTech firms. These guidelines dive into the nitty-gritty, providing recommended standards for the entire lifecycle of an AI model—from the data it's trained on to how it's monitored and updated over time. They emphasize the non-negotiable importance of reliability, fairness, robust risk management, and, above all, unwavering consumer protection.

Let's see how this works in the real world. Imagine you apply for a business loan through a major South Korean bank's mobile app. The decision, powered by AI, comes back in minutes: denied. Before these new regulations, that might have been the end of the road—a frustrating, opaque rejection. But under the AI Basic Act, the story is now completely different. The bank is legally obligated to provide you with a meaningful explanation. They can't just say "computer says no." They must disclose the main factors that led to the denial, such as, "The decision was heavily influenced by your company's short operating history and the high debt-to-income ratio detected in the data." Furthermore, they must give you a clear and simple way to appeal, to have a human loan officer review your entire application. This single change is revolutionary. It pierces the veil of the black box and empowers the individual, transforming them from a passive subject of the algorithm into an active participant with the right to an explanation and a second chance.

Part II: Balancing Innovation, Privacy, and Fairness

Crafting rules for AI is more than just writing laws; it's a delicate and precarious balancing act. South Korea's entire strategy can be visualized as a tightrope walker, carefully moving forward with a long pole. On one end of the pole is the immense weight of economic ambition and the drive for technological supremacy. On the other end is the equally heavy, non-negotiable commitment to protecting citizen privacy and ensuring fundamental fairness. Leaning too far in either direction could lead to a fall—either stifling the innovation that fuels the economy or betraying the public trust required for that economy to function.

On one hand, the government is flooring the accelerator. South Korea's national AI strategy is a bold declaration of its intent to be a global leader in the field. This isn't just talk; it's backed by massive public investment in computing infrastructure, like data centers and cloud networks, which are the essential building blocks for developing advanced AI. The government is also providing significant tax incentives and research grants to encourage private sector R&D, pouring fuel on the fire of an already vibrant tech industry. The message is clear: we want you to build, to experiment, and to win in the global tech race.

On the other hand, the government is diligently strengthening the guardrails. A key pillar of this effort is the recently revamped Personal Information Protection Act (PIPA). Updated in March 2025, the new PIPA is a powerful tool for consumer rights in the digital age. It grants individuals the right to data portability, meaning you can now demand your personal data from one company and transfer it to another in a machine-readable format. It's your data, and you have the right to take it with you.

More critically, PIPA directly confronts the challenge of automated decision-making. Under the new law, companies must inform you when a fully automated system, like an AI, is making a decision that significantly affects your rights. They are required to explain the criteria behind that decision and, crucially, provide you with a pathway to refuse the automated decision or demand a human review. This right is a powerful counterweight to the speed and scale of AI, ensuring that efficiency never comes at the cost of human dignity and oversight.

This entire balancing act is perfectly encapsulated in the philosophy of the AI Basic Act itself. While it sets clear and firm rules for high-impact systems, the penalties for breaking those rules—administrative fines up to about 30 million Korean Won—are noticeably modest. This is a stark contrast to regulations like the EU's AI Act, which threatens non-compliant companies with colossal, potentially business-ending fines. This difference is not an accident; it's a deliberate strategic choice. South Korea is signaling that it wants to use regulation as a guidepost, not a hammer. The goal is to foster a culture of responsible innovation through dialogue and guidance, rather than to create a climate of fear that might scare companies away from experimenting with bold new ideas.

Consider a final, practical example that brings this dual approach to life. A young customer uses a popular Korean FinTech app with a "robo-advisor" feature to manage their small investment portfolio. One day, the AI makes a series of trades that the customer doesn't understand. Under the new rules, they can exercise their right to an explanation for the automated decision and, if they remain concerned, demand that a human portfolio manager review the AI's strategy. This is the "protection" side of the balance in action. At the very same time, the FinTech company that created that robo-advisor may be benefiting from government tax credits for its AI research and using a state-supported cloud network to process its complex models. This is the "promotion" side. It is a perfect illustration of South Korea's grand experiment: empowering the consumer while simultaneously equipping the innovator, all in a bid to walk the tightrope into a fair and prosperous digital future.

A New Mandate for Banks and FinTechs: The Practical Realities of Compliance

The era of unchecked AI experimentation in South Korean finance is officially drawing to a close. For years, artificial intelligence was the exciting frontier, a kind of "Wild West" where data scientists and developers could build and deploy powerful models with few formal constraints. That frontier is now being settled. With the landmark AI Basic Act set to come into full force in January 2026, the clock is ticking. For every financial institution in the country, from the largest traditional banks to the most agile FinTech startups, this is a moment of profound change. The law signals a fundamental shift in mindset: AI is no longer a speculative side project run by the tech department; it is now a core business function that demands the same level of rigorous oversight, risk management, and executive accountability as any other critical operation.

This transformation requires moving AI out of the isolated research lab and into the formal structure of the boardroom. It means that the teams building credit scoring algorithms or fraud detection systems can no longer operate in a silo. Their work must be woven into the fabric of the organization's compliance and governance frameworks. The Chief Risk Officer, the Chief Compliance Officer, and even the CEO must now be literate in the basics of AI, capable of understanding its potential and holding their teams accountable for its responsible implementation. This is a cultural shift as much as it is a technical one, demanding new skills, new processes, and a new level of seriousness across the entire enterprise. The core message of the new regulatory landscape is simple: if an algorithm is making decisions that impact people's lives, it is no longer just code—it is a direct extension of the institution's values and responsibilities.

As the 2026 deadline approaches, financial firms are facing a pre-flight checklist of critical tasks. First and foremost is the mandate to document their AI models, effectively opening the black box from the inside. This goes far beyond simply saving the source code. It means creating a comprehensive, living record for every high-impact model. This documentation must explain the business case for the AI, the exact sources of data used for training and why they were chosen, and a detailed account of the testing and validation processes. It must also transparently state the model's known limitations and potential weaknesses. Think of it as creating a detailed pilot's manual for an algorithm. In the event of an audit or a critical failure, regulators and internal teams alike must be able to open this manual and understand exactly how the machine works, what it was designed to do, and where its blind spots are.

Next on the list is the imperative to conduct regular fairness and bias audits. This is the ethical stress test for every algorithm that holds a customer's financial future in its hands. It is no longer enough to simply build a model that is statistically accurate; it must also be demonstrably fair.

This involves a proactive and ongoing process of testing the model's outcomes across different demographic groups. Do loan approval rates differ significantly for men and women, even when all other financial factors are equal? Does the algorithm produce different results for applicants from different neighborhoods or age groups? Answering these questions requires specialized tools and dedicated teams, including both internal auditors and, increasingly, objective third-party experts. These audits are the financial world's equivalent of a regular health check-up, designed to detect and treat the poison of bias before it can cause systemic harm.

Finally, firms must build robust and user-friendly procedures for transparency and consumer redress. This is about creating the customer help desk for the AI age. It is a practical and human-centered challenge. It means training customer service representatives so they don't freeze up when a customer asks why an algorithm denied their credit card application. It means designing mobile apps and websites with clear, simple interfaces where users can request an explanation for an automated decision with a single click. Most importantly, it means establishing a well-defined and efficient process for escalating cases to a human reviewer. This cannot be a bureaucratic dead end; it must be a genuine pathway for appeal. Building this infrastructure is a significant investment, but it is the ultimate foundation for earning and maintaining public trust in an increasingly automated world.

Part II: The "Seoul Model": A Global Blueprint for the Algorithmic Age?

As South Korea charts its course, the world is watching. In capital cities and tech hubs from Brussels to Washington D.C. and Singapore, policymakers are wrestling with the exact same dilemma: how do you regulate a technology that is evolving at lightning speed without strangling the innovation that promises to drive future prosperity? The global conversation around AI governance has often been presented as a stark choice between two opposing models.

On one side, you have the European Union's approach, which can be seen as building a fortress of rights and regulations. With sweeping legislation like the GDPR and the highly comprehensive AI Act, the EU prioritizes safety, ethics, and fundamental rights above all else. Its approach is methodical, detailed, and backed by the threat of enormous, multi-billion-dollar fines. It is a model designed to provide maximum protection, even if it comes at the cost of slowing down the pace of innovation.

On the other side, you have the more traditionally market-driven approach of the United States, which has often resembled a dynamic frontier. The US has historically favored a sector-specific model, allowing industry to innovate freely and often stepping in with regulation only after a problem has become apparent. This approach fosters rapid experimentation and commercialization but can be slower to address systemic risks to consumers.

Into this landscape, South Korea's strategy emerges as a compelling "third way"—a pragmatic hybrid that some are beginning to call the "Seoul Model." It thoughtfully borrows elements from both worlds. From the EU, it takes the risk-based framework and the clear focus on protecting citizens' rights in high-impact areas. From the US, it takes the fierce, pro-innovation spirit and a preference for a more collaborative, less punitive enforcement style. It is an attempt to create a system that is both protective and pro-growth, a regulatory environment that functions not as a rigid cage but as a guiding trellis, providing structure for responsible growth.

For global policymakers, particularly those in other technologically advanced and highly competitive nations, this model is incredibly appealing. It demonstrates a pathway that avoids the extremes. It shows that it's possible to enforce meaningful ethical guardrails without bringing the machinery of innovation to a grinding halt. The emphasis on a national AI strategy, combining public investment and incentives with clear legal obligations, provides a holistic blueprint. The use of modest penalties suggests that the goal of regulation can be to foster a culture of compliance and continuous improvement, rather than simply to punish infractions. This "regulation as a guide" philosophy is especially attractive to nations that fear losing their competitive edge in the global tech race.

Of course, the Seoul Model is not a final, perfect answer. It is a bold, real-world experiment that is just beginning. Critical questions remain. Will the modest fines be a strong enough deterrent to ensure that all companies, big and small, take their obligations seriously? How will this carefully constructed framework adapt to the next wave of AI, such as the unpredictable and rapidly evolving power of generative AI? The answers to these questions will unfold in the coming years.

What is clear today, however, is that South Korea is offering more than just a set of new laws. It is offering the world a case study in pragmatic governance for the 21st century. It is a model built on the belief that public trust is the most valuable asset in a digital economy, and that long-term technological leadership cannot be sustained without a deep and abiding commitment to fairness.

France - Generative AI and Intellectual Property in Creative Industries

France and the Algorithmic Muse: Defending Droit d'Auteur in the Age of Generative AI

France, a nation where culture is not merely an industry but a cornerstone of national identity, finds itself at the epicenter of a technological earthquake. The ground is shaking beneath the foundations of its most cherished legal doctrine: the right of the author, or droit d'auteur. The source of this tremor is generative artificial intelligence, a powerful new force capable of creating art, music, and text with startling sophistication. This technology doesn't just challenge business models; it probes the very definition of creativity and authorship that France has championed for centuries. The ensuing conflict is not just a legal squabble. It is a profound philosophical debate about the future of human creation, with France acting as the primary defender of a deeply humanist legal tradition. The core of the French dilemma unfolds across two critical battlegrounds: the legality of using copyrighted works to train these powerful AI models, and the question of whether the resulting AI-generated output can ever be considered a "work" worthy of its own protection.

The Training Data Dilemma: A Question of Consent and Compensation

At the heart of every generative AI is a vast, invisible library. These models learn their craft by analyzing immense datasets of existing human creations—novels, paintings, songs, and photographs. In France, this process, known as Text and Data Mining (TDM), has ignited a firestorm. The legal framework is primarily shaped by France's implementation of the 2019 European Union

Copyright Directive. The law provides a specific exception for TDM, but it was designed with scientific research in mind, not for the creation of powerful commercial products that directly compete with the creators whose work was used for training.

The French system, unlike the American concept of "fair use," is not a flexible, open-ended standard. Instead, it relies on clearly defined exceptions. The TDM exception comes with a crucial string attached: an "opt-out" provision. This means that rights holders—the artists, writers, and musicians—can explicitly reserve their rights and forbid the use of their work for AI training. French creative guilds and collecting societies have seized upon this mechanism with vigor. Organizations representing authors, publishers, and visual artists are actively developing and promoting technical standards, such as machine-readable protocols, to make their opt-out declarations clear and unavoidable for any entity scraping data from the web.

This is more than a technical maneuver; it's a declaration of principle. From the French perspective, the uncompensated and non-consensual ingestion of their life's work is not a form of "fair" learning but a form of digital exploitation. They argue that tech companies are building multi-billion-dollar enterprises on a foundation of copyrighted material without permission or payment. The debate in France is therefore centered on re-establishing consent and creating viable licensing markets. The goal isn't necessarily to halt technological progress but to ensure that the value generated by AI flows back to the human creators who provided the essential raw material. This stance is creating significant friction, as AI developers argue that a fragmented, opt-out-driven landscape makes it impossible to build robust models, potentially putting European AI development at a global disadvantage.

The Ghost in the Machine: Can an Algorithm Be an Author?

The second great challenge posed by generative AI strikes at the philosophical core of French copyright law: the concept of authorship itself. Can a piece of music composed by an algorithm or an image conjured from a text prompt be granted copyright protection? Under the current French legal framework, the answer is a firm and resounding no. The tradition of droit d'auteur is inextricably linked to a human being. A work is protected because it is considered an extension of its creator's personality, bearing the unique l'empreinte de la personnalité de l'auteur—the imprint of the author's personality.

This principle demands that a work reflects the free and creative choices of a human mind. It is about the spark of individuality, the personal vision, and the intellectual effort that makes a work original. A machine, no matter how sophisticated, is not considered to possess a personality or the capacity for such creative expression. Therefore, a work generated entirely by an AI, with a human merely providing a simple prompt, would likely fall into the public domain in France. It lacks a human author in the sense that the law demands.

This creates a fascinating and complex legal gray area. What happens when a human artist uses AI as a sophisticated tool, extensively guiding, editing, and transforming the output? Here, the French courts would likely need to assess the level and nature of the human intervention. If the human's contribution is significant enough to impart their own creative personality onto the final product, then copyright could potentially be granted to the human author for that final, modified work. The legal test will not be about whether AI was used, but about whether a human author's creative choices were the decisive factor in the outcome. This maintains the humanist center of the law but opens the door to difficult case-by-case analyses that will have to distinguish between using AI as a paintbrush and treating it as the painter. The French system is thus bracing for a new wave of litigation that will force it to draw a line between human-assisted creation and autonomous machine generation.

The French Response: Courts, Committees, and Cultural Defense

France's response to the AI challenge has been swift and multi-pronged, unfolding in courtrooms, parliamentary committees, and high-level policy discussions. The nation's cultural institutions are not waiting passively. In a landmark move, French news agencies and publishers have been at the forefront of negotiating licensing deals with AI companies, leveraging a related legal concept known as "neighboring rights" to demand payment for the use of their content in services like AI-powered search. While not strictly about training generative models from scratch, these actions set a powerful precedent for valuing content and demanding compensation.

More direct legal challenges are also emerging. Groups of French visual artists and photographers have joined international lawsuits against AI companies like Midjourney and Stability AI, alleging massive copyright infringement through the unauthorized scraping of their online portfolios for training data. These cases, which are being watched closely across Europe, are the first major judicial tests of the TDM opt-out mechanism and the fundamental legality of the current AI development model. The outcomes will have profound implications for the balance of power between creators and tech giants.

Beyond the courts, French policymakers are actively studying the issue. The Conseil Supérieur de la Propriété Littéraire et Artistique (CSPLA), a key advisory body to the Ministry of Culture, has been tasked with producing in-depth reports on the impact of generative AI. These reports are meticulously analyzing questions of author remuneration, transparency in training data, and the potential need for new legislation. Discussions revolve around creating obligations for AI companies to maintain and disclose detailed records of the data used to train their models, which would empower rights holders to identify infringements and negotiate licenses. The overarching goal of these governmental efforts is to adapt the law without abandoning its core principles, ensuring that the French creative economy can innovate while still protecting the fundamental rights and dignity of its authors.

A Tale of Two Systems: The Humanist Fortress of Droit d'Auteur

The intensity of the French position is best understood when contrasted with the more flexible "fair use" doctrine found in the United States. Fair use is a defense against copyright infringement, evaluated on a case-by-case basis using four factors: the purpose of the use, the nature of the copyrighted work, the amount used, and the effect on the market. AI developers in the U.S. argue that training models is a "transformative" use—creating something new and different—and should therefore be considered fair use. This creates a vast, ambiguous legal terrain where the outcome is often uncertain until a court has ruled.

France rejects this ambiguity. Its droit d'auteur system is built not on a flexible defense but on a foundation of exclusive rights granted to the author. Any use of a work is considered infringement unless it falls under a specific, narrowly interpreted exception, such as the TDM exception with its opt-out. This approach provides creators with a much stronger and clearer starting position. It places the burden on the user—in this case, the AI company—to either secure a license or prove that their actions fit squarely within one of the limited legal exceptions.

This fundamental difference makes France a critical global case study. It represents a deliberate choice to prioritize the rights of the human creator over the unfettered freedom of technological development. The French debate is a powerful assertion that efficiency and innovation should not automatically override the moral and economic rights of authors. It is a defense of a cultural worldview that sees creative works not merely as data to be processed, but as vital expressions of human intellect and personality. As the world struggles to write the rules for the age of AI, France's steadfast defense of its humanist tradition offers a compelling, rights-holder-centric alternative to the more laissez-faire, tech-oriented approaches seen elsewhere. It is a high-stakes effort to ensure that the artists of tomorrow are not rendered obsolete by the very machines that learned from the masters of today.

The Training Dilemma: A Digital Feast on the Past

At the dawn of this new creative era stands a paradox. To build the future of art, music, and literature, generative AI models must first consume the past. They are voracious learners, and their classroom is the internet—a vast, sprawling library of human expression. These powerful systems are trained on datasets of a scale almost beyond human comprehension, datasets filled with the very books, photographs, paintings, and songs that are protected by copyright. This has ignited a fierce and fundamental conflict. The tech companies building these models see this process as a necessary step for innovation, a form of learning. But for the creators, artists, and authors of France, it feels like something else entirely: a grand, uncompensated appropriation of their life's work.

In the intricate world of French and European law, the battleground is a legal concept known as Text and Data Mining, or TDM. A carve-out in the law, born from the EU's Copyright Directive, allows for TDM, but it was designed primarily for the sterile environment of scientific research. For the bustling, high-stakes world of commercial AI, the rules are different. The law handed creators a shield: the right to "opt out." In theory, an artist or publisher can plant a digital flag on their work, a machine-readable notice that clearly states, "This is not for AI training." It was meant to be a simple assertion of consent, a way to control one's own creations in the digital wild.

But creators across France argue this shield is being treated as if it were invisible. They feel that AI developers are engaging in a form of digital trespass on a colossal scale, scraping and ingesting data first and leaving legal questions for later. The opt-out, they contend, is being willfully ignored. This frustration is no longer just a topic of debate in Parisian cafés or academic conferences; it has spilled into the courts. The lawsuit filed in March 2025 by a coalition of French authors and publishing houses against Meta is a thunderclap. It represents a direct challenge, a line drawn in the sand. The case, now before the Paris Judicial Court, is about more than just seeking financial damages. It is a demand for transparency. The plaintiffs want Meta to open the "black box" and reveal exactly which copyrighted works were used to train its AI. It is a foundational test of whether France's author-centric copyright laws can hold their ground against the immense momentum and resources of global technology giants.

To understand the human cost, consider the story of a single French photographer. Imagine she has spent decades honing her craft, capturing fleeting moments of light and life. Her personal website is her gallery, a carefully curated portfolio where each image is marked with a clear copyright notice. One day, she discovers that an AI image generator can replicate her unique aesthetic—her signature use of color, her compositional style—with a simple text prompt. Her work, her very artistic identity, has been deconstructed into mathematical patterns and absorbed by an algorithm. The AI company might argue that its scraping of her site was a legally permissible act of data mining. But for the photographer, it is an infringement not just of her reproduction rights, but of her artistic soul. This is the core of the French dilemma: a clash between a machine's need for data and a human being's fundamental right to control their own creation. The outcome of these legal battles will not just set a precedent; it will define the value of human creativity in a world increasingly filled with artificial ones.

The Output Problem: In Search of an Author

The second great question strikes at the very heart of what it means to create. When a generative AI, prompted by a user, produces a stunning image, a poignant poem, or a complex piece of music, who, if anyone, is the author? Can this new work be protected by copyright? In the French legal tradition, the answer has always been deeply personal. Copyright, or droit d'auteur, is not merely an economic right; it is a moral one. It springs into existence because a work is seen as an extension of a human creator's personality. It must bear what the law beautifully calls l'empreinte de la personnalité de l'auteur—the imprint of the author's personality. A work is protected because it carries a spark of its maker's soul, their unique choices, their vision, their struggles.

From this perspective, a work generated entirely by an algorithm has an empty space where its heart should be. A machine has no personality, no lived experiences, no emotions to channel into its work. Therefore, the prevailing view in France, as in much of the world, is that a purely AI-generated piece of content cannot be copyrighted. It lacks a human author and thus falls into the public domain, free for anyone to use. But technology is rarely so simple. The line becomes wonderfully, and maddeningly, blurry when a human collaborates deeply with an AI. What if an artist spends hours crafting intricate prompts, generating dozens of images, then meticulously selecting, editing, and combining elements from each to create a final masterpiece? Here, the human is not just a passive prompter; they are a director, a curator, a true artist using a revolutionary new tool. In these cases, French law would likely find that the artist's significant creative intervention has imprinted their personality onto the final work, making it eligible for copyright protection. The key is no longer the tool that was used, but the depth of the human hand that guided it.

This complex issue has prompted bold new thinking. A legislative proposal brought before the French National Assembly in September 2023 offered a radical and elegant solution. It sought to solve the output problem by tying it directly back to the input dilemma. The proposal suggested that for works created by AI without "direct human intervention," the copyright should not disappear into the public domain. Instead, it should be automatically assigned to the original creators whose works were used to train the AI in the first place. This is a revolutionary idea. It reframes AI generation not as a form of creation from nothing, but as a sophisticated form of derivation. It proposes a system of legal and economic justice, ensuring that the value created by the AI flows back to the human artists who provided its education.

Let's apply this to the real world. Imagine a famous living painter in Lyon, known for her vibrant, unmistakable style. An AI is trained on her entire body of work. A user then types a prompt: "Create a scene of a rainy Parisian street in the style of this Lyonnaise painter." The AI generates a beautiful image that is undeniably an echo of her genius. Under the 2023 proposal, the copyright for this new image would not belong to the user who wrote the simple prompt, nor would it belong to the AI company. It would be vested in the painter herself. This legal mechanism would protect her from having her style endlessly replicated without her consent or compensation. It is a profound statement that an artist's style—the very essence of their creative identity—is worthy of protection, even from the ghost in the machine. This French approach represents a powerful effort to re-center the human in the creative process, ensuring that technology serves art, and not the other way around.

France's Regulatory and Policy Response: Building the Guardrails

As the tidal wave of generative AI reshapes the creative landscape, France is not standing passively on the shore. Instead, it is actively and methodically building a system of guardrails and channels, attempting to guide the powerful currents of technology. The French approach is not about building a dam to stop innovation. It is about constructing a sophisticated irrigation system, one designed to ensure that the benefits of this new technological river nourish the entire creative ecosystem, rather than flooding it. This effort is a complex tapestry woven from high-level European directives, focused national policies, and the diligent work of specialized regulatory bodies.

At the very top, the guiding light is the European Union's landmark AI Act, a sweeping piece of legislation that officially came into force in August 2024. For creators, its most crucial element is a single, powerful mandate: transparency. The Act requires developers of the most powerful AI models to pull back the curtain on their training processes. They are now legally obligated to publish a "sufficiently detailed summary" of the copyrighted content they have used. This rule is a direct response to the "black box" problem, where creators knew their work was being used but had no way to prove it. The transparency obligation is designed to be a tool of empowerment, giving artists, authors, and publishers the information they need to stand up and exercise their rights, including the fundamental right to opt out of being used for commercial data mining.

Back in France, a dedicated orchestra of institutions is working to turn this broad European mandate into a national reality. Each plays a distinct but harmonious part. The Generative AI Committee, established in the autumn of 2023, acts as the conductor, bringing together experts from the worlds of culture, technology, and economics to advise the government. Its mission is to chart a strategic course, ensuring France can be a leader in the AI revolution without sacrificing its cultural soul. Playing alongside is the CNIL, France's formidable data protection authority. While its primary focus is on protecting the private data of citizens, its work is crucial. The CNIL pushes for strong governance, ensuring that personal information isn't carelessly swept up in the vast data trawls used for AI training, and that individuals are properly informed when their data is being used.

Perhaps the most vital voice for creators in this arrangement is the CSPLA, the High Council for Literary and Artistic Property. In a pivotal report published in December 2024, the CSPLA took the AI Act's abstract phrase—"sufficiently detailed summary"—and gave it sharp, practical meaning. The council understood that for transparency to be effective, it had to be meaningful. They clarified that a vague, top-level summary would not suffice. An AI company simply stating that it trained its model on content from "Wikipedia.org" or "the internet" would be failing its duty. The CSPLA's vision is far more granular. It proposed that companies should provide information in a way that is genuinely useful to a rights holder, perhaps through a searchable online database or by establishing a dedicated contact point. This would open a door for real dialogue, transforming a potential legal battle into a possible licensing negotiation. It is a quintessentially French solution: pragmatic, detailed, and designed to protect the rights of the individual creator above all else.

The Path Forward: A Delicate Balance Between Two Futures

The road ahead for France is as promising as it is perilous. The legal and legislative frameworks are still being poured, and the concrete is far from dry. The ambitious legislative proposal introduced in 2023 stands as a key blueprint for the future, yet it is also a source of intense debate. While praised for its vision, critics worry that its bold ideas might prove too cumbersome in practice, potentially creating a labyrinth of legal obligations that could hinder the very innovation it seeks to manage. The document is less of a finished law and more of a national statement of intent—a declaration that as AI evolves, the core principles of author remuneration and creative control must evolve with it.

At the heart of this proposed future are two foundational pillars: radical transparency and fair compensation. The first pillar is the call for a mandatory "AI-generated work" label. This simple requirement for disclosure would have profound effects. It respects the audience, allowing consumers of media, art, and information to know the origin of what they are seeing. It also provides an essential tool for creators and researchers to track the spread and influence of AI-generated content on our shared culture. The second, more complex pillar is the system for ensuring creators are paid. The proposal reaffirms the audacious idea that in the absence of significant human intervention, the copyright of an AI's output should belong to the owners of the works it was trained on. This is then coupled with a mandate for fair remuneration for any creator whose work is used in an AI's education.

To solve the immense puzzle of how to actually distribute this remuneration, the proposal looks to a familiar model: the Collective Management Organization, or CMO. In the worlds of music and publishing, CMOs are the trusted intermediaries that collect royalties and ensure they find their way to the right artists and authors. Applying this system to the scale of AI is both brilliantly logical and almost impossibly difficult. The challenge is one of sheer magnitude. A modern AI model may be trained on billions of data points—a fragment of a sentence from one author, a sliver of a photograph from another. The task of tracking every source and attributing its precise influence on a final output is, as many have noted, a monumental one. It's like trying to thank every drop of rain that helped a forest grow.

This is the core challenge that defines the path forward for France. How can the nation adapt its proud, human-centric legal tradition to a new technological paradigm? The goal is to strike an incredibly delicate balance. On one side is the urgent need to foster innovation and remain competitive in a field that will define the global economy for the next century. On the other is an unwavering commitment to the principles of droit d'auteur—the belief that human creativity is unique, valuable, and deserving of protection and reward. The French solution will not be simple. It will require new technologies for tracking data, new models for licensing content, and a new spirit of collaboration between the worlds of art and engineering. France is walking a tightrope, and the world is watching to see if it can successfully build a bridge to a future where artificial intelligence and human ingenuity do not just coexist, but thrive together.

Canada - AI and Environmental Sustainability: The Hidden Costs

Introduction: A Nation at a Crossroads

Canada stands at a fascinating, yet precarious, intersection. On one hand, the nation is celebrated globally as a pioneer in artificial intelligence. Fueled by the groundbreaking work of researchers like Geoffrey Hinton and Yoshua Bengio, Canada has meticulously crafted its image as a thoughtful leader in the AI space. The Pan-Canadian AI Strategy is more than just a policy document; it's a declaration of ambition, positioning the country as a hub for ethical, human-centric AI development. Yet, beneath this glossy surface of innovation lies a deep and troubling contradiction. As we race to build a digital future, we are tethered to a physical world with finite resources. The very machines that promise to solve our most complex problems are creating a voracious and often invisible demand for energy, water, and raw materials.

This is the hidden cost of the AI revolution, a cost that runs directly counter to Canada's other identity: that of a nation with vast, pristine wilderness and ambitious climate commitments. The tension is palpable. How can a country that champions both technological supremacy and environmental stewardship reconcile the two? The computational engine of AI, from the sprawling data centers humming with activity to the complex global supply chains that produce its hardware, casts a long and growing environmental shadow. This analysis delves into that shadow, exploring the collision between Canada's digital dreams and its ecological realities. We are forced to ask a difficult question: Is our pursuit of artificial intelligence sustainable, or are we building a brilliant future on an unstable environmental foundation?

The Unseen Price: Energy, Water, and Waste

The engine of modern AI is the data center, and these facilities are incredibly thirsty. Their thirst is primarily for two resources: electricity and water. Training a single large-scale AI model, like the ones that power popular generative AI tools, can consume an astonishing amount of energy. Estimates suggest it can be equivalent to the annual electricity consumption of hundreds of North American homes. In Canada, this presents a complex picture. While provinces like Quebec and British Columbia boast vast hydroelectric power reserves, attracting data centers with the promise of "clean" energy, this narrative isn't as simple as it seems. Using this green energy to power AI means it isn't being used for other essential purposes, like electrifying transportation or heating homes, potentially forcing other sectors to rely on fossil fuels. In provinces with more carbon-intensive grids, like Alberta and Ontario, the direct carbon footprint of AI computation is even more pronounced.

Beyond electricity, there is the often-overlooked issue of water. Data centers generate immense heat, and water is the most common method for cooling the thousands of servers packed inside. These systems can consume millions of liters of water per day, a fact that is becoming increasingly alarming in an era of climate change-induced droughts and water scarcity. This water is drawn from local municipal supplies or natural bodies, putting a strain on ecosystems and communities. The water is often evaporated in cooling towers and cannot be returned to its source, representing a net loss to the local watershed.

The environmental impact doesn't stop at operations. The lifecycle of AI hardware carries its own heavy burden. Manufacturing the sophisticated semiconductors and servers required for AI is a resource-intensive process. It relies on the mining of rare-earth minerals, often in environmentally damaging ways, and involves complex global supply chains with significant carbon footprints. Then there is the end of the line: electronic waste, or e-waste. AI development moves at a blistering pace, rendering powerful hardware obsolete in just a few years. This cycle of rapid replacement generates mountains of discarded electronics, filled with toxic materials that can leach into soil and groundwater if not disposed of properly. For Canada, a nation built on its natural wealth, this full-spectrum environmental cost of AI presents a profound and urgent challenge.

The Policy Paradox: Fueling the Fire While Fighting It

Canada's approach to managing AI's environmental impact is a study in contradiction. The federal government is essentially walking in two opposite directions at the same time, creating a fascinating and risky policy paradox. On one side, it is pouring billions of dollars into accelerating the very industry that is driving up resource consumption. The recently announced Canadian Sovereign AI Compute Strategy, backed by a $2 billion investment, is a clear statement of intent. The goal is to build massive, state-of-the-art computing infrastructure to ensure Canada remains at the forefront of AI development, preventing a "compute drain" to other countries. This strategy is driven by powerful economic incentives; hosting energy-hungry data centers is big business, and being a leader in AI is seen as critical for future prosperity. This policy aggressively subsidizes the problem: the massive and growing energy and resource footprint of the AI sector.

Simultaneously, the government is also championing "Green AI." It funds research and promotes projects that use artificial intelligence as a tool to solve our most pressing environmental problems. Canadian researchers are using AI to model climate change with greater accuracy, predict the path of devastating wildfires, optimize energy grids for efficiency, and monitor endangered species. These initiatives are genuinely promising, showcasing AI's incredible potential as a force for good. They align perfectly with Canada's global reputation for environmental stewardship and its commitments under international climate agreements like the Paris Accord.

Herein lies the paradox. The government is using taxpayer money to both subsidize the expansion of AI's environmental footprint and fund the development of AI-based solutions to clean it up. It is like trying to bail out a sinking boat while simultaneously drilling more holes in the hull. This dual approach reveals a deep-seated conflict within national priorities. It reflects a desire to have it all—economic growth driven by a high-tech industry and a sustainable, green future—without fully grappling with the reality that, under the current model, these two goals are often in direct opposition. This creates a policy landscape where the problem and the proposed solution are locked in an escalating, and perhaps unsustainable, arms race against each other.

Forging a Sustainable Path: Ethics and Long-Term Vision

To navigate this complex dilemma, we must turn to deeper principles, namely those found in Environmental Ethics and the concept of Sustainable Development. Environmental ethics pushes us beyond a human-centered view of the world. It asks us to consider the moral standing of the non-human world—the rivers, forests, and ecosystems that are impacted by our technological ambitions. From this perspective, is the development of a slightly more capable AI model a justifiable reason to divert millions of liters of water from a stressed watershed or to generate a carbon footprint equivalent to a small town? It forces a moral calculus that weighs the tangible, immediate harm to ecosystems against the often abstract or future benefits of technological progress. This ethical framework challenges the very idea that progress at any environmental cost is acceptable.

The concept of Sustainable Development provides a more pragmatic, yet equally powerful, lens. The famous definition—meeting the needs of the present without compromising the ability of future generations to meet their own needs—serves as a crucial benchmark. When we apply this to Canada's current AI trajectory, serious questions arise. Is the explosive growth in energy and water consumption for AI truly meeting a "need" of the present, or is it fueling a want for technological supremacy and corporate profit? More importantly, is this resource depletion compromising the ability of future Canadians to have access to stable energy grids, sufficient clean water, and a stable climate? The current model, which prioritizes computational power above all else, appears to be fundamentally at odds with this intergenerational contract.

Finding a way forward requires a deliberate and courageous shift in policy and practice. Canada must move beyond its paradoxical stance and forge an integrated strategy. This could involve implementing stringent "green computing" standards for any data center receiving public funds, ensuring they meet the highest levels of energy and water efficiency. It means demanding radical transparency from tech companies about the environmental cost of their AI models. It also means investing heavily in research not just for bigger AI, but for better AI—more efficient algorithms that can achieve incredible results with a fraction of the computational power. This path requires reimagining AI development not as a race for infinite scale, but as a discipline of optimization and elegance, where the goal is to achieve the most intelligence with the least possible environmental impact. It is about choosing a future where our digital creations enrich our world without impoverishing the planet that sustains us all.

Conclusion: A Choice for the Future

Canada's journey with artificial intelligence is emblematic of a global challenge. The nation has successfully positioned itself as a world leader in AI research and ethical considerations, but it has yet to fully confront the profound environmental consequences of its own success. The hidden costs—the immense consumption of energy and water, the creation of e-waste, and the carbon footprint of hardware—are accumulating, creating a significant and unacknowledged debt against our climate goals. The current policy approach, a paradoxical strategy of simultaneously subsidizing the problem and funding its potential solutions, is not a sustainable long-term plan. It represents a deep-seated reluctance to make the hard choices that genuine sustainability requires.

Ultimately, the conflict between Canada's AI ambitions and its environmental commitments is not a technical problem to be solved by a clever new algorithm. It is a question of values. It forces us to decide what kind of future we are building. Will we pursue technological advancement at any cost, accepting ecological degradation as the price of progress? Or will we have the wisdom to guide the development of AI in a way that respects planetary boundaries and honors our responsibility to future generations? Canada, with its unique position as both an AI powerhouse and a nation defined by its natural heritage, has a critical opportunity to lead. By moving beyond its current contradictions and pioneering a truly sustainable approach to artificial intelligence, Canada can prove that the brightest technological future is one that is in harmony with the natural world.

The Environmental Footprint of AI: A Debt to the Physical World

We interact with artificial intelligence as if it were magic. We speak a question into the air, and an answer materializes on our screen. We type a simple prompt, and a stunning, original image blooms into existence. This digital world feels clean, weightless, and ethereal. But this is an illusion. Behind every seamless AI interaction, there is a vast, hidden, and profoundly physical world of machinery. This machinery lives in sprawling, windowless buildings called data centers, and it is hungry. It has a voracious appetite for electricity, an unquenchable thirst for water, and a lifecycle that begins deep within the earth and ends on a growing mountain of electronic waste. The true cost of our digital magic is a debt being charged to the planet, and the bill is coming due.

The Thirst for Power: AI and Energy Consumption

The fundamental process of artificial intelligence is computation. It is an endless series of mathematical calculations performed at blinding speed. Every one of these calculations, no matter how small, requires a tiny spark of electricity. When you scale this up to the level of a modern AI model, with trillions of parameters being adjusted millions of times over, those tiny sparks combine into a raging inferno of energy demand. The data centers that house these operations are not just server rooms; they are industrial-scale power consumers.

Think of the process of "training" an AI model. This is where the machine learns by analyzing colossal amounts of data. It is an incredibly intense, one-time effort that can last for weeks or even months. During this period, thousands of specialized computer chips run at maximum capacity, 24 hours a day. The energy consumed is staggering. The example that training one model can emit as much carbon as five cars over their entire lifetimes is a startling starting point. But let's unpack that. It's not just about the fuel; it's the manufacturing of those cars, all their maintenance, and every kilometer they will ever drive. An AI model can create that much environmental impact before it even answers a single question for a user. The training of Google's BERT model, a foundational piece of technology, consumed enough electricity to power over 100 average Canadian homes for an entire year.

This colossal energy demand is only set to explode. As AI becomes more integrated into our daily lives, the global electricity consumption of data centers is on a terrifying trajectory. The projected leap from 460 terawatt-hours in 2022 to nearly 1,050 by 2026 is difficult to comprehend. To put that 1,050 terawatt-hour figure in perspective, it is more than the current annual electricity consumption of India, a nation of 1.4 billion people. We are effectively building a new, continent-sized industrial power user dedicated solely to computation.

The source of this energy is the critical factor. In places with carbon-intensive energy grids, every AI query directly contributes to greenhouse gas emissions. Even in places like Canada, which leverages significant hydropower, this massive demand creates an "opportunity cost." The clean energy being funneled into data centers is energy that cannot be used to decarbonize other sectors, like transportation or heavy industry. The digital revolution, powered by AI, is inadvertently creating one of the largest and fastest-growing energy challenges of the 21st century.

The Unseen Coolant: AI's Water Footprint

Energy consumption is only half of the story. Where there is immense electrical power, there is immense heat. The thousands of processors inside a data center, all running at full tilt, generate a level of heat that would quickly melt them into slag if left unchecked. To prevent this, data centers require constant, powerful cooling, and the most common and efficient method for this is using water. This gives rise to AI's second hidden environmental cost: its massive water footprint.

Large data centers are essentially giant evaporative coolers. They pull in vast quantities of water from local sources—rivers, lakes, or municipal supplies—and circulate it to absorb the heat from the servers. This heated water is then often pumped to cooling towers, where it is exposed to the air and evaporates, carrying the heat away into the atmosphere. The key issue is that this water is consumed; it is turned into vapor and is not returned to the local watershed. A single large data center can consume millions of liters of water every single day, equivalent to the daily water needs of a small city.

The projected global water footprint of AI data centers—up to 6.6 billion cubic meters annually by 2027—is a truly colossal figure. This is more than one-third of the total annual water consumption of the United Kingdom. It is enough water to fill over 2.6 million Olympic-sized swimming pools. This demand places an enormous strain on local ecosystems and communities, especially in a world increasingly plagued by drought and water scarcity due to climate change. We are seeing a future where communities may be asked to conserve water and let their fields go fallow, while just down the road, a data center is evaporating millions of liters a day to generate AI images or chat responses.

This creates profound ethical dilemmas. Is it right to build these water-guzzling facilities in regions that are already water-stressed? The water footprint doesn't even stop there. If the electricity powering the data center comes from thermoelectric sources like coal, natural gas, or nuclear plants, those power plants themselves require enormous amounts of water for their own cooling systems. This creates a compounding effect, where AI's thirst for electricity indirectly amplifies its thirst for water. The clean, digital interface of AI completely masks this messy, wet, and resource-intensive reality.

From Mine to Landfill: The Hardware and E-Waste Crisis

The environmental impact of AI begins long before a single watt of electricity is used, and it continues long after the servers are shut down. The physical hardware—the silicon heart of the machine—carries a heavy environmental price tag at every stage of its life. This is the third, and perhaps most tangible, hidden cost of artificial intelligence.

The journey of an AI chip, like a Graphics Processing Unit (GPU), begins deep inside the earth. Its creation requires a cocktail of materials, including silicon, copper, gold, and a host of rare-earth minerals. The mining and refining of these materials are incredibly destructive processes, often leading to habitat destruction, soil erosion, and water contamination. These raw materials are then shipped around the world in a complex, carbon-intensive supply chain to highly specialized fabrication plants. The process of turning silicon wafers into functional microchips is one of the most resource-intensive manufacturing processes on the planet, requiring immense amounts of purified water and stable electricity.

The life of this sophisticated hardware is brutal and short. The race for AI supremacy has created a culture of rapid obsolescence. AI researchers and companies demand the fastest, most powerful chips available, and the technology is advancing so quickly that a state-of-the-art GPU can be considered outdated in just two or three years. This isn't because the old hardware is broken; it's simply not competitive enough. This relentless upgrade cycle fuels a growing tsunami of electronic waste, or e-waste.

E-waste is not like ordinary trash. These discarded servers, processors, and circuit boards are laden with toxic materials like lead, mercury, and cadmium. If they end up in a landfill, these toxins can leach into the soil and contaminate groundwater for generations. While some e-waste is recycled, the process is complex and often incomplete. Much of it is shipped to developing nations, where informal recycling operations can expose workers, including children, to hazardous fumes and materials with little to no protection. We are essentially exporting the toxic legacy of our technological progress. The pursuit of a non-physical intelligence is, ironically, creating a very physical and very poisonous waste problem for the entire planet.

The Great Canadian AI Gamble: Building the Engine and the Antidote

Canada finds itself in a peculiar and defining moment, navigating a future it is actively helping to create. On one hand, the nation is pouring billions of dollars into constructing the raw, physical infrastructure of the artificial intelligence age—a vast and power-hungry network of data centers and supercomputers. On the other, it is championing AI as a uniquely powerful tool to solve the very environmental crises this new infrastructure threatens to worsen. This is Canada's great AI gamble: a high-stakes bet that the solutions born from artificial intelligence will be potent enough to heal the wounds inflicted by the machine itself. It is a story of immense ambition, profound contradiction, and a search for a delicate balance between building the future and preserving the planet.

Forging the Digital Iron: Canada's Push for AI Supremacy

Canada's journey into AI began with a bold declaration. In 2017, it became the first country in the world to launch a national AI strategy. The Pan-Canadian AI Strategy was a visionary move, a $125 million investment designed to attract and retain top-tier research talent, cementing the nation's reputation as a global brain trust for artificial intelligence. This initial phase was about planting the flag, establishing Canada as a place where the foundational ideas of the next technological revolution were born and nurtured. The strategy was a resounding success, creating a vibrant ecosystem of research and innovation.

But ideas alone do not power an AI revolution. The second phase of the strategy recognized a critical truth: modern AI runs on computational horsepower. The most brilliant algorithms are useless without access to massive-scale computing infrastructure. This realization led to the landmark announcement in 2024 of the Canadian Sovereign AI Compute Strategy. This was not a subtle adjustment; it was a seismic shift in commitment, backed by an enormous $2 billion investment. The message was clear: Canada was no longer content to just be AI's research lab; it intended to be one of its primary engine rooms.

This strategy is about building the digital equivalent of a national railway system. It's about ensuring that Canadian researchers and businesses are not forced to take their world-changing ideas elsewhere to find the computing power needed to bring them to life. The plan is twofold. First, it involves building public supercomputing facilities, creating a national resource for the most ambitious research projects. Second, and perhaps more significantly, it aims to mobilize a torrent of private sector investment through initiatives like the "AI Compute Challenge." This challenge, supported by up to $700 million, is essentially a call to the world's tech giants, inviting them to build their next generation of commercial AI data centers on Canadian soil.

Here, however, the strategy confronts the unavoidable environmental elephant in the room. The government knows that data centers are voracious consumers of energy and water. To square this circle, Canada is playing its trump card: its abundance of clean energy and its cool climate. A proposed $15 billion investment fund is designed to specifically incentivize the development of "green" AI data centers. The AI Compute Challenge itself explicitly seeks proposals for "innovative and sustainable compute solutions." The pitch is compelling: come build your power-hungry AI infrastructure here, where the grid is cleaner and the cold northern air provides natural, energy-saving cooling. It is an attempt to build a resource-intensive industry in the most resource-efficient way possible. Yet, it remains a fundamental contradiction—aggressively courting an industry known for its massive environmental footprint while simultaneously trying to brand the effort as a green initiative.

The Digital Solution: AI as a Force for Climate Action

While one arm of Canadian policy is busy building the physical home for AI, the other is passionately deploying AI as a frontline soldier in the war against climate change. This is the other side of the gamble—the belief that for all its environmental costs, AI can provide solutions that are exponentially more valuable. Across the country, in labs, forests, and cities, this belief is being put into practice in remarkable ways.

In the heart of Canada's natural resource economy, AI is transforming how we manage the nation's vast forests. Instead of relying solely on manual surveys, AI-powered drones and satellites scan millions of hectares from above. Their algorithms can detect the subtle discoloration of leaves that signals the start of a pine beetle infestation, allowing for targeted interventions before an entire forest is lost. In an era of escalating wildfires, predictive AI models are becoming indispensable. They analyze a complex web of data—humidity levels, wind patterns, vegetation density, and lightning strike data—to identify high-risk areas and simulate how a fire might spread, giving emergency crews a critical head start.

This same intelligence is being applied to revolutionize our energy systems. Making the switch to renewables like wind and solar comes with a challenge: intermittency. The wind doesn't always blow, and the sun doesn't always shine. Calgary-based WindESCo is a prime example of how AI can tackle this. Their AI platform acts as a smart conductor for an entire wind farm. It analyzes real-time weather data and grid demand, then makes constant, tiny adjustments to the pitch and yaw of each individual turbine blade. It's a level of micro-optimization no human team could ever achieve, and it results in a tangible increase in energy production—up to 3% more clean power from the exact same hardware.

AI is also helping Canadians adapt to the climate changes that are already here. In cities like Calgary, the "AI for the Resilient City" tool, supported by Microsoft, is a powerful new weapon against urban heat islands. The AI analyzes satellite imagery and local demographic data to pinpoint neighborhoods that are most vulnerable to extreme heat waves. It can show city planners exactly where a new park or a stand of trees would have the most cooling impact, protecting public health in a warming world. This work extends into pure environmental science. At the IISD Experimental Lakes Area in Ontario, decades of meticulous records on when the lake ice breaks up each spring—a key climate change indicator—have been fed into a machine learning model. The model can now predict the "ice-off" date with incredible accuracy, helping scientists better plan their crucial field research seasons.

Perhaps most poignantly, Canadian innovation is using AI to help us breathe. As wildfire seasons worsen, the threat of smoke is no longer a distant concern but an immediate public health crisis. Researchers at Carleton University have developed a sophisticated AI model that can simulate how toxic wildfire smoke particles infiltrate and spread within indoor spaces. This isn't just an academic exercise; this tool can directly inform architects and engineers, helping them design buildings with smarter ventilation systems that can protect occupants during a smoke event. It is a direct, tangible application of AI to help humanity adapt to a harsher, climate-altered reality.

The Defining Choice: Weaving Sustainability into the Code of AI

Canada is writing a complex and fascinating story about its future. It is a tale of two ambitions, running parallel yet constantly threatening to collide. In one lane, the nation is flooring the accelerator on building a massive, powerful, and resource-intensive AI infrastructure. In the other, it is carefully cultivating AI as a delicate, intelligent tool to solve the world's most pressing environmental problems. This dual strategy is more than just a policy quirk; it is the stage for a profound ethical dilemma. We are forced to confront a series of difficult and uncomfortable questions about the true cost of progress. The path forward requires more than just innovation; it demands wisdom, foresight, and a fundamental rethinking of what it means to build responsible technology in a world of finite limits.

The Ethical Calculus: Balancing Progress with Planetary Boundaries

At the heart of Canada's AI strategy lies a difficult and often unasked question: what is the net environmental impact of this technology? We are captivated by AI's potential for good. An AI that optimizes a city's power grid could prevent tons of carbon from being emitted. An AI that designs a more efficient solar panel could accelerate our transition to clean energy. But these benefits do not exist in a vacuum. The very AI model that optimizes the power grid may have required the energy equivalent of a small town to train. The supercomputer that designed the new solar panel may have consumed millions of liters of water for cooling. This is the ethical tightrope Canada is walking.

We must begin to perform a new kind of environmental accounting, a "carbon calculus" for every major AI initiative. Does the good outweigh the bad? Can the insights from an AI-driven climate model, which might help us predict sea-level rise, truly justify the construction of a new data center on land that could have been a wetland, absorbing carbon and filtering water? There are no easy answers. The benefits are often diffuse and long-term, while the costs—in energy, water, and materials—are concrete, immediate, and growing exponentially. To navigate this dilemma, we cannot simply hope for the best. We must actively build a new framework for AI governance, one that hardwires environmental sustainability into the entire lifecycle of the technology.

This begins with a radical shift in how we create AI. For years, the mantra in AI development has been "bigger is better." This has led to an arms race to build ever-larger models, regardless of the computational cost. The new frontier must be a race for efficiency. We must promote and fund research into smaller, more elegant algorithms and specialized hardware that can achieve incredible results with a fraction of the power. This is the philosophy of "less is more" applied to the digital age.

Next, we must demand radical transparency. The environmental footprint of a major AI model should not be a corporate secret. We need a mandatory reporting system, akin to a nutritional label on food. Companies deploying large-scale AI should be required to disclose its energy consumption, its water usage, and its estimated carbon footprint. This information is critical. It allows policymakers, and the public, to make informed decisions. Canada's existing frameworks for responsible AI, like the Digital Charter, which champion principles of accountability and transparency, provide a perfect foundation. We simply need to expand their scope to include environmental accountability.

Of course, the infrastructure itself must be green. Canada's push to power new data centers with renewable energy is a vital and necessary step. But it is not a cure-all. We must also prioritize technologies that dramatically reduce the need for water in cooling, and we must locate these facilities in a way that minimizes their impact on local ecosystems. Finally, we must address the ghost in the machine: e-waste. We need to foster a circular economy for AI hardware. This means designing servers and chips not to be disposable, but to be durable, repairable, and easily recyclable. It means turning the end of one product's life into the beginning of another's.

This is not just a theoretical exercise. There are concrete examples of this new direction taking shape. The Government of Canada's investment in the "AI Pathways" project, led by the Alberta Machine Intelligence Institute (Amii), is a brilliant case study. This $9 million initiative is dedicated to training thousands of workers from the traditional energy sector in AI skills, specifically to help them transition into careers in the clean energy economy. It is a direct and powerful linkage, using AI not just as a tool for environmental analysis, but as a bridge for people, helping to ensure that the transition to a greener future is a just and inclusive one. It is a sign that we can, if we choose, align our technological ambitions with our environmental obligations.

From Theory to Action: The Practical Implications

The ethical dilemmas surrounding AI and the environment are no longer abstract, future-tense problems. They demand immediate and practical action from every corner of society. The choices we make today will fundamentally shape the world we inhabit for decades to come.

For Canada's policymakers, the message is clear: environmental considerations can no longer be an afterthought in national AI strategies. They must be a core, non-negotiable component from the very beginning. It is not enough to simply fund innovation and hope that it turns out to be green. Governments must become active gardeners, carefully cultivating the kind of AI ecosystem they want to see grow. This means using the full suite of policy tools available. It means offering targeted investments and powerful tax incentives for companies that pioneer green computing. It means implementing clear, mandatory standards for environmental reporting. It means embedding environmental impact assessments into the approval process for any new large-scale data center or government-funded AI project. Just as we have building codes to ensure our physical structures are safe, we need a "green code" for our digital infrastructure to ensure it is sustainable.

For the technology companies, researchers, and engineers who are actually building this future, there is a profound ethical imperative to act. The culture of AI development must evolve. For too long, the primary metrics of success have been accuracy and speed. A model was considered "better" if it was more powerful and produced more precise results, with little to no regard for the resources consumed in the process. This must change. Resource efficiency must be elevated to a first-class metric, standing shoulder-to-shoulder with accuracy. The brightest minds at institutions like Mila in Montreal or the Vector Institute in Toronto should be challenged not just to build the most powerful AI, but to build the most efficient AI. The new moonshot should not be a model with ten trillion parameters, but a model that can perform complex tasks on a simple laptop. This is a challenge worthy of our best innovators.

Ultimately, this entire conversation must be framed by the concept of planetary boundaries. Scientists have identified several critical Earth systems—such as climate stability, freshwater availability, and biodiversity—that have fundamental limits. Pushing past these boundaries could trigger irreversible and catastrophic changes to the world as we know it. The unbridled, exponential growth of AI's environmental footprint represents a real and present danger of accelerating our collision with these limits. The pursuit of artificial intelligence, a technology of seemingly infinite possibility, cannot be decoupled from the reality of the finite planet on which it operates.

The true test of a "responsible AI" is not merely that it is fair, unbiased, and transparent. The ultimate test is whether it contributes to the flourishing of life on Earth, or to its diminishment. Canada stands at a crossroads. It can continue down its paradoxical path, fueling a problem with one hand while trying to solve it with the other. Or, it can make a courageous choice. It can decide to lead the world in pioneering a new model of technological development—one that is not just intelligent, but wise. A future where our most advanced tools are used not to conquer nature, but to find our place within it.

Australia - Indigenous Data Sovereignty and AI

Introduction: A New Frontier for an Ancient Sovereignty

We live in an age powered by data. It has become the invisible current that shapes our world, influencing everything from the news we read to the services we receive. In this new world, artificial intelligence (AI) is the master navigator, using vast oceans of data to learn, predict, and act. For many, this is a story of progress and innovation. But for the First Peoples of Australia, Aboriginal and Torres Strait Islander peoples, this new digital landscape feels eerily familiar. It presents a new frontier where old battles for sovereignty, for the right to self-determination, must be fought once again.

This is not just about technology. It's about stories, identity, and power. The intersection of AI and Indigenous Data Sovereignty in Australia is a critical crossroads. Down one path lies the potential for a new wave of digital colonization, where Indigenous data is taken and used without consent, perpetuating cycles of harm and misunderstanding. On the other path, however, lies an incredible opportunity. It's a future where Indigenous communities themselves are in control, harnessing these powerful new tools to strengthen culture, revitalize languages, and care for Country in ways that blend ancient wisdom with cutting-edge technology. This chapter explores that very crossroads, examining the profound risks and the burgeoning hope that AI brings to the struggle for Indigenous digital self-determination.

The Perils: Data Colonization and the Algorithmic Echo Chamber

When AI systems are built without Indigenous oversight, they don't just create new problems; they often amplify old ones. The most significant risk is a modern form of colonization: data colonization. For generations, data about Aboriginal and Torres Strait Islander peoples was extracted by governments, researchers, and institutions. It was collected about them, but rarely for or with them. This data often painted a picture of deficit, focusing on problems rather than strengths, and was used to justify policies that caused deep harm. Today, this historical data forms the training ground for many AI systems.

When an AI is trained on biased data, it learns to be biased itself. This is called algorithmic bias. Imagine an AI system designed to predict who might need extra support from social services. If it's trained on historical data that unfairly targeted Indigenous families for intervention, the AI will learn that pattern. It will then start flagging Indigenous families at a higher rate, not because of genuine risk, but because it is simply echoing the prejudices of the past. It becomes a high-tech echo chamber for systemic racism, making it harder for people to get fair access to housing, loans, or even justice. The algorithm, cloaked in the language of objective technology, gives a new, dangerous life to old injustices.

Furthermore, there is the profound risk to Indigenous Cultural and Intellectual Property (ICIP). ICIP includes everything from traditional stories and art to knowledge about medicinal plants and land management practices. This knowledge is collectively owned and passed down through generations. AI systems, designed to scrape and learn from vast amounts of information, have no understanding of these sacred protocols. An AI could potentially hoover up sacred knowledge from the internet, misappropriate it, or even generate new art in a traditional style without any benefit or permission from the community who are its rightful custodians. Without Indigenous governance, AI threatens to turn sacred cultural heritage into just another data point, stripped of its context, spirit, and meaning.

The Promise: Indigenous-Led AI for Culture and Country

Despite the serious risks, the story of AI in Indigenous Australia is not solely one of peril. Across the continent, a powerful counter-narrative is emerging. Aboriginal and Torres Strait Islander innovators, technologists, and communities are not just resisting data colonization; they are actively seizing AI as a tool for their own empowerment. They are proving that when technology is guided by Indigenous values and governance, it can become a powerful force for cultural survival and revitalization.

One of the most exciting frontiers is language preservation. Many of Australia's hundreds of original languages are critically endangered. Traditional methods of preservation are vital, but technology offers a new toolkit. Indigenous-led initiatives are now exploring AI to build interactive language learning apps. These apps can be tailored to specific local dialects, using AI to help with pronunciation and create engaging learning experiences for young people. It's a way to bring ancient words into the digital world where the next generation lives, ensuring the songs and stories of their ancestors continue to be spoken.

Another remarkable area is caring for Country. Indigenous peoples have been managing their lands sustainably for over 60,000 years, accumulating an incredibly sophisticated body of ecological knowledge. Today,

Indigenous ranger groups are combining this traditional knowledge with AI-powered tools. They use drone imagery and AI analysis to monitor the health of land and sea Country, track the movements of animals, identify invasive species, and manage cultural burning practices more effectively. The AI doesn't replace traditional knowledge; it serves it. It processes vast amounts of environmental data, providing insights that help rangers make better decisions, guided by a cultural responsibility to care for the land. In these hands, AI becomes a tool not of extraction, but of stewardship and connection. These Indigenous-led projects are rewriting the rules, demonstrating a future where technology serves community goals and strengthens cultural identity.

The Framework: Building a Sovereign Digital Future

The struggle to control the narrative and the data has moved from community projects to the national policy stage. At the heart of this movement are the principles of Indigenous Data Sovereignty. This is the simple yet profound assertion that Indigenous peoples have the right to govern their own data. It's about recognizing that data derived from Indigenous peoples is not an open resource but a critical asset that belongs to them. This right to govern includes how data is collected, who can access it, how it is used, and how it is stored. It's about ensuring that data is used to benefit Indigenous communities, not to harm them.

These ideas have been crystallized into internationally recognized frameworks, such as the CARE Principles for Indigenous Data Governance. CARE is an acronym that stands for Collective benefit, Authority to control, Responsibility, and Ethics. These principles demand that data ecosystems be designed to empower, not exploit. They shift the focus from a top-down, extractive model to one based on partnership, respect, and self-determination. It's a call to move beyond simply asking for consent to a system where Indigenous peoples have genuine authority and control over their digital heritage.

In Australia, these principles are starting to shape government policy. A landmark development has been the work towards a National Framework for the Governance of Indigenous Data. This initiative, co-designed with Indigenous leaders and organizations, represents a crucial step towards embedding Indigenous Data Sovereignty into the machinery of government. The framework aims to create clear rules for how government agencies collect and use data about Aboriginal and Torres Strait Islander peoples. It seeks to ensure that data collection is purpose-driven, community-led, and ultimately, accountable to Indigenous Australians. While the journey is far from over, this framework is a vital case study. It demonstrates a nation grappling with its colonial past and attempting to build a more equitable digital future, providing a potential blueprint for how states can begin to respect and legislate for the digital self-determination of First Peoples in the age of AI.

Conclusion: The Digital Songline of Self-Determination

The dawn of the AI era has placed Australia's First Peoples at a familiar historical juncture. Just as they have fought for sovereignty over their lands, waters, and cultures, they now fight for sovereignty over their data. The challenges are immense. The threat of algorithmic bias entrenching disadvantage and the risk of digital platforms misappropriating sacred knowledge are profoundly real. The ghosts of past colonial practices loom large, warning of a future where technological progress for some comes at the cost of justice for others.

Yet, to only see the risks is to miss half the story. The true power of this moment lies in the resilience, innovation, and leadership being shown by Aboriginal and Torres Strait Islander peoples. They are not waiting to be given a seat at the table; they are building their own. By embedding ancient values into the design of new technologies, they are creating AI that serves community, revitalizes language, and helps heal Country. They are forging a new path, demonstrating that technology does not have to be a colonizing force. It can be a tool for liberation.

The struggle for Indigenous Data Sovereignty in the age of AI is, ultimately, the next chapter in the long and enduring story of self-determination. It is about ensuring that Indigenous Australians are not just subjects of the data revolution, but are its architects, its storytellers, and its navigators. It's about charting a new digital songline, one where the future is not just predicted by an algorithm, but is actively and sovereignly chosen by the people themselves.

Theoretical Underpinnings

More Than Tech: A Battle for Power and a Postcolonial Lens

To truly understand the conversation around Indigenous data and artificial intelligence, we must first agree on what we are talking about. And it's not just about technology. It isn't a simple debate over software, servers, or algorithms. At its heart, this is a deeply political issue, a modern chapter in a very old story about power, control, and the right to exist on one's own terms. To make sense of it, we need to look through the lens of Postcolonial Theory. This way of thinking helps us see the long shadows that colonial empires still cast upon our modern world. It reveals how the power imbalances of the past haven't disappeared; they have simply changed their shape.

Postcolonialism teaches us that the end of an empire doesn't mean the end of colonial structures. These structures live on in our laws, our institutions, and even in the way we think about knowledge itself. The debate over who controls Indigenous data is a perfect example. It is a direct continuation of historical struggles for sovereignty over land and culture. For centuries, the narrative of Indigenous peoples was written by outsiders—explorers, anthropologists, and government officials who came, observed, and took. Today, the tools are different. They are databases, algorithms, and vast digital networks. But the fundamental question of power remains the same: Who gets to tell the story? Who gets to benefit from the knowledge? Framing the issue this way moves it out of the IT department and into the realm of human rights and decolonization. It's not a technical problem to be solved; it is a colonial legacy to be dismantled.

Digital Colonialism: The New Scramble for Data Territory

Building on this postcolonial understanding, we arrive at a powerful and unsettling concept: digital colonialism. Think back to the historical colonization of Australia. Ships arrived, flags were planted, and land was claimed and exploited for resources that would enrich a distant empire. Digital colonialism follows the exact same pattern, but the territory is intangible, and the resource is data. Today, the ships are the massive tech corporations, research institutions, and even well-meaning government agencies that sweep across the digital landscape. They arrive with their own tools and their own worldview, planting digital flags as they extract vast quantities of information.

This data—stories, health records, languages, biometric information, ecological knowledge—is then transported to servers far away. It is processed and refined, creating value in the form of insights, commercial products, or government policies. Just like the resources of the past, this new wealth rarely flows back to the communities from which the raw material was taken. Instead, it serves the interests of the colonizing power. This process replicates the harms of the past. It renders Indigenous peoples as passive subjects to be studied, rather than active agents of their own destiny. It takes their knowledge, their identities, and their stories, and turns them into a resource for others to mine. Digital colonialism is the quiet, insidious process by which the sovereignty of a people can be eroded, one dataset at a time, without a single shot being fired. It is the new frontier of an old exploitation.

Indigenous Data Sovereignty: A Declaration of Digital Self-Determination

In response to this rising tide of digital colonialism, a powerful global movement has emerged: Indigenous Data Sovereignty (IDSov). This is more than a policy recommendation; it is a profound assertion of inherent right. IDSov declares that Indigenous peoples have the right to govern their own data at every stage of its life. This includes the right to control how their data is collected, who has ownership of it, and how it is ultimately used and shared. It is the digital expression of the fundamental right to self-determination. It is the people themselves saying, "Our stories, our knowledge, our information—they belong to us."

This movement is not happening in a vacuum. It is deeply rooted in international human rights law, particularly the United Nations Declaration on the Rights of Indigenous Peoples (UNDRIP). The UNDRIP affirms the rights of

Indigenous peoples to maintain and control their cultural heritage and traditional knowledge. Indigenous Data Sovereignty simply extends this established right into the digital realm. It makes the case that in the 21st century, control over information is just as critical to cultural survival and political autonomy as control over land. IDSov is therefore the political and ethical framework for decolonizing data. It provides the language and the principles for Indigenous nations to challenge extractive practices and to begin building their own data ecosystems—systems designed by them, for them, and grounded in their own unique values and laws. It is a declaration of independence for the information age.

A Clash of Two Worlds: The Commodity Versus the Connection

The tension at the heart of the debate between AI and Indigenous data comes from a fundamental clash between two entirely different ways of seeing the world. It is a collision of worldviews. On one side, we have the dominant Western approach, which has become the default operating system for the entire digital economy. In this view, data is essentially a commodity. It is seen as a raw material, like oil or timber, that can be extracted, refined, and sold. Information is often depersonalized, stripped of its original context, and treated as a universal, objective fact. The more data you can collect, the better, because its value is realized when it is amassed in huge quantities and processed for patterns. This is the logic that powers big data and artificial intelligence.

On the other side stands an Indigenous epistemology—a different way of knowing and being. From this perspective, data and knowledge are not inanimate commodities. They are alive. They are relational, contextual, and often sacred. Information isn't an isolated "fact"; it is part of a web of relationships that connects people to each other, to their ancestors, and to Country. "Country" itself is not just a patch of land; it is a living system of which humans are an interconnected part. Therefore, a piece of data—be it a traditional story, the location of a sacred site, or a family's health information—is never just a data point. It carries with it responsibilities, obligations, and a spirit. It belongs not to an individual, but to the collective. This worldview does not ask, "What can we extract from this information?" It asks, "What is our responsibility to this knowledge?"

AI's Design Flaw: The Hunger for Decontextualized Data

This is where artificial intelligence enters the picture, and where the clash of worldviews becomes a direct technical and ethical problem. AI systems, particularly the machine learning models that dominate the field today, are designed to work almost exclusively within the Western, extractivist paradigm. Their very nature is to ingest absolutely massive, often decontextualized datasets and search for patterns. An AI does not understand relationships. It does not understand responsibility. It does not understand the sacred. It only understands patterns in numbers.

To an AI, a sacred story transcribed onto a webpage is no different from a weather report. A genealogical record connecting a family to their ancestral lands is no different from a customer's shopping history. The system is built on the premise of stripping away context to find a universal, mathematical signal within the noise. This inherent mismatch is the root cause of the immense ethical risks AI poses to Indigenous communities. When you feed an AI data that is deeply contextual and relational, the AI, by its very design, ignores that context. It breaks the relationships, violates the protocols, and flattens sacred knowledge into a simple, usable commodity. The system is not intentionally malicious; it is simply designed with a worldview that is fundamentally incompatible with an Indigenous understanding of knowledge. This makes standard AI development an act of profound cultural violence unless it is deliberately and radically reshaped.

The True Struggle: Forcing Technology to Respect an Indigenous Worldview

Therefore, the fight for Indigenous Data Sovereignty is not a fight against technology itself. It is a struggle to redefine it. It is an effort to force the world of AI development to bend and accommodate an Indigenous worldview, rather than the other way around. This is a demand to move beyond the shallow ethics of "do no harm" and towards a deeper ethics of relationality and responsibility. It is a call to build new systems from the ground up, guided by Indigenous principles.

This means demanding that technology respect the core tenets of consent, context, and collective benefit. Consent can no longer be a one-time checkbox on a form; it must be a continuous, meaningful dialogue with the collective. Context can no longer be stripped away; it must be preserved and encoded alongside the data itself, ensuring knowledge is never separated from its meaning and its source. And most importantly, the application of any technology must be for the collective benefit of the community, as defined by the community itself. This is a radical re-imagining of what technology can be. It is a struggle to create AI that is not just intelligent in processing patterns, but wise in understanding relationships. It is the work of decolonization in action, ensuring that the future is not just coded by a powerful few, but is co-designed with the wisdom of the world's oldest living cultures.

The Risks of AI to Indigenous Peoples and Cultures

When new technologies are rolled out, they are often presented as neutral tools, promising a future of efficiency and progress for everyone. But for Aboriginal and Torres Strait Islander communities, history has taught a hard lesson: without their control and oversight, new tools can simply become new weapons to inflict old harms. Artificial intelligence, for all its potential, is no exception. When developed and deployed from a purely Western, technical perspective, AI systems can become powerful engines for injustice, perpetuating the legacies of colonization in a new and insidious digital form. The risks are not theoretical; they are tangible threats to culture, livelihood, and the fundamental right to self-determination.

Coded Prejudice: Algorithmic Bias and Discrimination

At the very core of modern AI is the data it learns from. An AI is like a student; it only knows what it is taught. If its textbooks are biased, the student will become biased too. This is the problem of algorithmic bias. The vast majority of data used to train AI systems today is drawn from mainstream society. This data often completely excludes or deeply misrepresents Australia's First Peoples. The digital world has rendered them invisible or, worse, visible only through a lens of disadvantage created by centuries of systemic discrimination.

Now, imagine this biased AI is put to work in a critical area like the justice system. An algorithm might be designed to assess a person's risk of reoffending. If it has been trained on historical police data—data that reflects a long history of over-policing and unfairly targeting Indigenous communities—the AI will learn a flawed and racist lesson. It will learn to associate being Indigenous with being high-risk. The algorithm doesn't understand the context of colonization, intergenerational trauma, or systemic racism. It just sees a pattern in the data and follows it blindly. The result? An Indigenous person could be denied bail or receive a harsher sentence, not because of their actions, but because a biased algorithm gave their case a "high-risk" flag. The decision is laundered through the language of objective technology, making the prejudice harder to see and even harder to challenge. This isn't a future possibility; it is a present-day danger, creating a high-tech pipeline that reinforces the very discrimination our society claims to be fighting against.

Digital Forgery: The Misappropriation of Culture

Perhaps the most immediate and visceral threat comes from generative AI—systems that can create new content, like images and text. For Indigenous culture, this technology poses a profound threat to Indigenous Cultural and Intellectual Property (ICIP). ICIP is not like Western copyright; it is a living, collective inheritance of art, stories, songs, and sacred knowledge, governed by complex protocols and responsibilities passed down through generations. Generative AI has no respect for these protocols. It is designed to inhale everything it can find on the internet, learn the patterns, and then create new things that mimic the original.

This can lead to a devastating form of digital appropriation. An AI can be trained on thousands of images of sacred rock art or traditional dot paintings, scraped from museum websites and online galleries without the free, prior, and informed consent of the communities whose ancestors created them. The AI can then spit out an endless stream of "new" images in a so-called "Aboriginal style." These creations are hollow forgeries. They are culturally meaningless, stripped of the story, the Law, and the connection to Country that gives authentic Indigenous art its power and meaning.

This causes deep cultural harm by flooding the world with fake, soulless imagery that misrepresents and homogenizes hundreds of distinct and diverse Aboriginal and Torres Strait Islander cultures. It also causes direct economic harm. An authentic Indigenous artist spends years, even a lifetime, learning their craft and the stories they have the right to tell. They are now forced to compete in an online marketplace against algorithms that can generate thousands of cheap imitations in a matter of seconds. As a stark example, in early 2025, the creative world was rocked by criticism directed at Adobe's massive stock image library. The platform was found to be hosting a flood of AI-generated images that were falsely and offensively labeled with tags like "Indigenous Australian" and "Aboriginal artwork." These images were a disturbing mash-up of generic symbols, inaccurate designs, and nonsensical body markings. They were a digital caricature of a living culture, vividly illustrating how generative AI can become a tool for mass-producing cultural disrespect and devaluing the work of real Indigenous creators.

Digital Terra Nullius: The New Wave of Data Colonization

Underpinning all these risks is the overarching threat of data colonization. The legal fiction of terra nullius—the false idea that Australia was an empty land belonging to no one before British settlement—was used to justify the colonization of the continent. Today, we are witnessing the rise of a "digital terra nullius." This is the dangerous assumption that the digital realm, and the data within it, is an empty frontier, free for the taking. Tech companies, governments, and researchers often act as if Indigenous data is an unclaimed resource, ignoring the fact that this data is an extension of Indigenous peoples and is subject to their own governance and laws.

This unchecked collection of data, often for use in opaque AI systems, is a modern form of dispossession. Data about communities can be gathered for one stated purpose, like a census, and then repurposed for something else entirely, such as surveillance or predictive policing, without the community's knowledge or consent. This information can be used to build systems that do not benefit, and may actively harm, the very people from whom it was taken. It creates a profound power imbalance, where the people who know the most about Indigenous communities are not Indigenous people themselves, but distant bureaucrats or corporations. This digital land grab silences Indigenous voices and undermines their sovereignty, treating their digital identities as territory to be claimed and controlled by others.

The Principles of Indigenous Data Sovereignty (IDSov)

In the face of these immense challenges, Indigenous peoples are not standing by as passive victims of technological change. They are responding with a powerful, proactive, and globally coordinated movement to reclaim control. This response is built on the principles of Indigenous Data Sovereignty (IDSov). In Australia, organizations like the Maiam nayri Wingara (which translates to 'good path' or 'the right way' in the Ngunnawal language group) Aboriginal and Torres Strait Islander Data Sovereignty Collective have been at the forefront, articulating a clear vision for a decolonized digital future. The core assertion of IDSov is simple and non-negotiable: Indigenous peoples have the inherent right to govern their data across its entire lifecycle, from its creation to its application and its eventual destruction. This is about flipping the power dynamic from one of extraction to one of self-determination.

To put this vision into practice, a clear set of guidelines was needed. This led to the development of the CARE Principles for Indigenous Data Governance. The CARE principles were created to sit alongside the more common technical data principles known as FAIR (Findable, Accessible, Interoperable, Reusable). While FAIR principles are valuable for making data useful to machines and researchers, they say nothing about people or ethics. The CARE principles fill this critical gap, centering human rights and Indigenous worldviews. They provide a clear framework for anyone working with Indigenous data to ensure their work is empowering, not exploitative.

The Four Pillars of CARE

The CARE principles are built on four interconnected pillars:

Collective Benefit: This is the first and most important question. Who does this work actually help? The principle of Collective Benefit demands that data collection and any resulting technology, including AI, must be designed to directly benefit Indigenous communities. The value generated from the data should flow back to the people it came from. This means moving away from "research for research's sake" and towards projects that address community-defined needs and priorities, whether that's improving health outcomes, revitalizing language, or managing Country. It ensures that communities are not just research subjects, but are the primary beneficiaries of any knowledge created from their information.

Authority to Control: This principle directly confronts the legacy of colonialism. For generations, others have held the authority to tell Indigenous stories and control

Indigenous information. Authority to Control asserts that Indigenous peoples themselves have the right to make decisions about their data. This includes having control over who can access it, what it can be used for, and how it is governed. It means recognizing Indigenous governance structures and laws as being legitimate and central to any project. This is the practical expression of self-determination; it's about putting the remote control for the data firmly in the hands of the community.

Responsibility: This pillar places a clear obligation on those who work with Indigenous data. They have a responsibility not just to the data itself, but to the people it represents. This means being transparent about how the data is being used and for what purpose. It means building genuine, long-term relationships based on trust and respect, not just transactional, short-term projects. It also means actively working to build data capabilities within Indigenous communities, ensuring they have the skills and resources to manage their own data infrastructure. It's about being a good partner, not just a data collector.

Ethics: Finally, the principle of Ethics grounds the entire framework. It insists that the rights, well-being, and worldview of Indigenous peoples must be the primary concern at every single stage of the data lifecycle. This goes far beyond a simple ethics committee approval. It means considering the potential for harm from an Indigenous perspective, understanding the deep connections between data, land, and spirit, and ensuring that the process respects cultural protocols. It is a commitment to ensuring that the pursuit of knowledge never comes at the cost of a people's dignity or sovereignty.

To see how this works, imagine a research group wants to use AI to analyze health data from a specific Aboriginal community to understand chronic disease. Following the CARE principles, their first step would not be to apply for a grant, but to go to the community and build a relationship. The project would be co-designed from day one. Community Elders would help form a governance council (Authority to Control). The project's goals would be set by the community to address their most pressing health concerns (Collective Benefit). The researchers would be transparent about their methods and would train local community members in data skills (Responsibility). And every step, from data collection to the final AI analysis, would be guided by protocols established by the community to ensure it aligns with their values (Ethics). The result is not just better science; it is ethical science that empowers the community it is meant to serve.

The Australian Government's Response: The GID Framework

For generations, the relationship between Indigenous Australians and the government has been defined by a one-way flow of information. Data was extracted from communities, held in government databases, and used to make policies for Aboriginal and Torres Strait Islander peoples, but rarely with them. This created a deep and justified mistrust. It was a system that treated people as subjects to be managed rather than as partners in their own future. But the powerful and persistent advocacy of Indigenous leaders and data experts has begun to turn this great ship of state. There is a growing recognition within government that the old ways have failed and that a new relationship, built on respect and shared power, is the only path forward.

This recognition has crystallized into a landmark policy. After a long and intensive process of co-design with Aboriginal and Torres Strait Islander partners, the Australian

Government, led by the National Indigenous Australians Agency (NIAA), officially launched the Framework for Governance of Indigenous Data (GID) in 2024. This is not just another report destined to sit on a shelf. The GID Framework is a foundational piece of policy architecture designed to fundamentally change how the entire Australian Public Service (APS) works. Its mission is to take the principles of Indigenous Data Sovereignty—the right of Indigenous peoples to control their own data—and embed them into the daily practices and culture of government. It is an official acknowledgment that Indigenous data is not a government asset, but a treasure that belongs to the people from whom it is derived.

A New Charter for Data Relationships

The GID Framework is built on four deceptively simple, yet revolutionary, principles that are intended to guide every government interaction involving Indigenous data.

First, Partner with Aboriginal and Torres Strait Islander people. This is the cornerstone of the entire framework. It signals a move away from the tokenistic "consultation" of the past towards genuine, enduring partnerships. It means Indigenous representatives are at the decision-making table from the very beginning of a project, not just invited to comment at the end. They are co-designers, co-creators, and co-owners of the process.

Second, Build data-related capabilities. This principle recognizes that power and skill go hand in hand. For too long, the technical expertise for data analysis has been held exclusively within government and research institutions. This principle commits the government to not only train its own staff to work in culturally appropriate ways but, crucially, to invest in building the data skills and technological resources within Indigenous communities and organizations. The goal is to create a future where communities can manage, analyze, and use their own data for their own purposes.

Third, Provide knowledge of data assets. This is a direct response to the "black box" of government data holdings. For decades, communities had little to no idea what information the government held about them or how it was being used. This principle mandates transparency. It requires agencies to be open about their data, making it clear what they collect, why they collect it, and who has access to it. It is about restoring the right of Indigenous people to see and understand their own digital reflection as it is held by the state.

Fourth, Build an inclusive data system. This is a commitment to redesigning the very infrastructure of government data. It means creating systems that can accommodate and reflect Indigenous worldviews. This could mean finding ways to represent kinship structures, cultural geography, or collective ownership—concepts that don't fit neatly into standard spreadsheets. It is about rebuilding the digital house so that it is welcoming and functional for everyone.

To ensure these principles are more than just aspirations, the framework carries real weight. Starting in January 2025, every single Australian Government agency is required to begin developing a detailed, seven-year implementation plan to show exactly how they will align their practices with the GID Framework. This work is directly tied to the commitments made in the National Agreement on Closing the Gap, which identifies shared access to data and transforming government organizations as essential for achieving real change.

To see the difference this makes, consider an agency wanting to use AI to understand trends in Indigenous education. The old way would be for analysts in Canberra to take national datasets, run them through an algorithm, and produce a report highlighting deficits. Under the GID Framework, the agency would be required to form a partnership with Indigenous education experts and community leaders first. This group would decide what questions the AI should even be asking, ensuring a focus on strengths and success. They would help interpret the data, providing the cultural context the algorithm lacks, and would co-design how the insights are used to create community-led education solutions. It's a seismic shift from a top-down, extractive process to a collaborative journey of discovery.

Indigenous-Led AI: Opportunities for Cultural Revitalization

While holding governments and institutions accountable is one half of the story, the other, more exciting half is about what Indigenous communities are building for themselves. They are not just focused on mitigating the risks of AI; they are actively seizing it as a powerful tool for cultural survival and self-determination. By taking the lead in designing and developing AI systems, Aboriginal and Torres Strait Islander innovators are embedding their own knowledge, values, and protocols directly into the technology. They are proving that AI does not have to be a colonizing force. When built by the community, for the community, it can become a powerful ally in the work of strengthening culture.

Breathing New Life into Ancient Languages

Of the hundreds of distinct languages spoken in Australia at the time of invasion, the vast majority are now critically endangered. The voices of Elders who hold this linguistic knowledge are fading. AI offers a new and powerful toolkit in the race against time. Indigenous-led projects are now using advanced speech recognition and natural language processing to document and revitalize these languages. This is so much more than just recording words. Technologists are working side-by-side with Elders to build AI models that can understand the unique sounds and grammatical structures of a specific language. This technology can then be used to create interactive learning apps for young people, to transcribe and translate precious historical recordings, or even to power tools that allow speakers to text and communicate online in their own tongue. When these projects are Indigenous-led, they ensure that sacred stories are handled with the correct protocols and that the technology serves the ultimate goal of putting the language back into the mouths of the community.

Caring for Country in the 21st Century

Indigenous peoples have been sustainably managing their land and sea Country for millennia, guided by a deep and holistic body of traditional ecological knowledge. Today, this ancient wisdom is being combined with cutting-edge AI. Indigenous ranger groups, who are on the front lines of conservation, are using AI as a partner in their work. They use drones to survey vast and often remote areas of land. The AI then analyzes this imagery, capable of identifying things the human eye might miss—the early spread of an invasive weed, changes in water quality, or the health of specific vegetation. This allows rangers to work more efficiently and effectively. It doesn't replace their knowledge; it enhances it. They combine the AI's data with their own understanding of the seasons, the animals, and the songlines to make the best decisions for healing and managing their Country.

A stunning example of this is happening in Kakadu National Park. Traditional Owners are working with scientists from Australia's national science agency, CSIRO, to tackle para grass, an invasive weed that chokes the wetlands and devastates local ecosystems. An AI-driven software analyzes high-resolution drone footage to precisely map the location and density of the weed. This information gives Indigenous rangers a clear picture of where to focus their management efforts. The results have been remarkable. By controlling the para grass, they have seen a massive resurgence in the population of magpie geese, a vital traditional food source and a key indicator that the Country is healthy and thriving. Critically, the project's data is governed by the community's rules. It is structured in "three rings" of access. The outer ring, containing general maps, is public. The middle ring, with more detailed operational data, is for the rangers. But the inner ring, holding the most sensitive ecological and cultural knowledge, is accessible only to Traditional Owners and Elders. This is a masterful example of forcing technology to conform to ancient cultural protocols, using AI as a sharp and powerful tool to fulfill a sacred responsibility to care for Country.

Telling the Old Stories in New Ways

The heart of culture is storytelling. Indigenous creators are now exploring how AI, augmented reality (AR), and other digital tools can carry their ancient narratives into the future. They are creating immersive experiences that allow stories to be seen and felt, not just heard. Imagine a young person walking on Country, holding up a phone or tablet, and using an AR app to see the creation story of that very place unfold before their eyes. Imagine an AI helping to animate traditional art, bringing the characters and symbols to life to share their meaning with a new generation. By putting these tools in the hands of Indigenous artists, they are ensuring their stories continue to evolve and resonate, finding new expression in the digital songlines of the 21st century.

Practical Implications

The rise of artificial intelligence and the powerful counter-movement for Indigenous Data Sovereignty are not abstract, academic debates. They are creating a new reality on the ground in Australia, a reality that comes with profound and urgent practical implications for everyone involved. The theoretical clashes and policy frameworks discussed previously now demand real-world changes in how people work, think, and relate to one another. For the nation to navigate this new technological frontier ethically and justly, three key groups—government policymakers, technology creators, and Indigenous communities themselves—must each embrace a fundamental and challenging transformation. This is the practical work of decolonizing the digital future.

For Policymakers and Government Agencies: The End of Business as Usual

For the Australian Public Service, the launch of the Framework for Governance of Indigenous Data (GID) is not an update; it is a system reboot. It signals the definitive end of the colonial-era model of governance, where decisions were made in centralized offices and imposed upon First Nations peoples. The practical implications are stark: the familiar, top-down processes of the past are no longer acceptable. The new baseline is genuine partnership and the explicit sharing of power.

Moving beyond "consultation" is the first and most critical hurdle. For decades, consultation has often been a shallow, tick-a-box exercise—a series of meetings where government officials present a pre-determined plan and ask for feedback. This is not partnership. Genuine partnership, as demanded by the GID Framework, means Aboriginal and Torres Strait Islander people are in the room with real authority from the very beginning. In practical terms, this means co-designing projects from a blank slate. It means Indigenous experts and community representatives sitting on steering committees with equal voting rights. It means co-drafting the terms of reference for any data project, including those involving AI. It means sharing control over budgets and resources, ensuring that funding flows to community-controlled organizations to lead the work themselves.

This requires a profound cultural shift within government agencies. The mindset must change from "we are the experts" to "we hold one form of expertise, and the community holds the essential expertise of lived experience, culture, and context." Public servants need new skills. They need training in cultural safety, in the history of colonization, and in the principles of co-design. They need to learn to listen more than they speak. They must become facilitators and allies, not just administrators.

When it comes to AI, this new way of working is non-negotiable. An agency cannot ethically develop an AI system that will impact Indigenous lives without a community-led governance body overseeing the entire project. This body must have the power to approve the datasets used for training, to audit the algorithm for bias, and, most importantly, to veto the project if it is deemed to be against the community's interests. This is what power-sharing looks like in practice. It is messy, it takes more time, and it requires a humility that has not always been a feature of government. But it is the only way to build trust and to ensure that the powerful tools of the state are used to serve Indigenous aspirations, not to perpetuate old injustices. The GID Framework is the new map, and partnership is the only legitimate way to travel.

For Technology Developers and Researchers: A New Ethical Compass

For the innovators—the coders, the data scientists, the AI researchers, and the tech companies—the era of consequence-free creation is over. The "move fast and break things" ethos of Silicon Valley is dangerously incompatible with the responsibilities that come with handling Indigenous data. Working in this space requires a radical expansion of the technologist's toolkit, adding a new set of ethical and practical competencies that are just as important as knowing how to write code.

The first and most fundamental commitment must be to upholding the CARE Principles for Indigenous Data Governance. Before a single line of code is written, a developer must ask themselves: Who will collectively benefit from this technology? Do the people the data comes from have genuine authority to control its use? Am I, the creator, accepting my responsibility to be transparent and build capacity? Is this project ethical from an Indigenous worldview, considering its potential impact on culture, kinship, and well-being? These questions must become as integral to the development process as debugging and testing.

This leads directly to the necessity of deep and respectful collaboration. Technologists must see themselves not as lone creators building a product, but as partners in a community-led endeavor. This means slowing down. It means investing significant time in building relationships on the ground, long before a project is even defined. It means learning the basics of local cultural protocols, understanding the history of the community, and recognizing that trust is the most valuable commodity. It is a process that cannot be rushed or managed through Zoom calls alone.

Practically, this new approach involves using new tools designed specifically for this ethical landscape. Among the most important are Traditional Knowledge (TK) and Biocultural (BC) Labels. These are digital tags that can be embedded into a file's metadata. Think of them as a permanent "care label" for a piece of information. A TK Label can clearly state the rules for using a piece of knowledge—for example, a "TK attribution" label ensures the original community is always credited, while a "TK sacred" label can warn that the information is secret and not for public view. A BC Label can show that a piece of scientific data about a plant is inextricably linked to the biocultural knowledge of a specific Indigenous group. These labels make Indigenous protocols machine-readable, ensuring that the rules for use travel with the data wherever it goes. Using these labels is a tangible way for a researcher or developer to show they respect Indigenous governance and are building systems that can understand and enforce it. The challenge for the tech world is to move from a mindset of extraction to one of reciprocity, building technology that doesn't just take, but gives back, guided by a new ethical compass.

For Indigenous Communities: Building a Sovereign Digital Nation

For Aboriginal and Torres Strait Islander peoples, the rapid advance of AI is both a threat and an opportunity of historic proportions. It presents an urgent, undeniable imperative to act decisively to secure the digital future. This is the work of modern nation-building. It involves a concerted effort to build digital capabilities from within and to proactively assert data sovereignty, ensuring that technology becomes a tool for liberation and the fulfillment of long-held aspirations.

The first practical step is building a digitally skilled workforce. This goes far beyond basic computer literacy. It means nurturing the next generation of Indigenous data scientists who can analyze their own data, AI ethicists who can design culturally-grounded governance frameworks, software developers who can build community-owned applications, and digital storytellers who can preserve culture in new mediums. This requires investment in culturally appropriate STEM education, creating pathways into technology careers, and establishing Indigenous-led tech hubs and innovation centers. The goal is not just to have a seat at the table, but to build a new table altogether.

Alongside building human capacity, there is the work of building sovereign digital infrastructure. This means developing community-owned and controlled data storage solutions—sometimes called "digital keeping places" or data trusts. Instead of community data being held on government or corporate servers, it can be stored in a digital repository governed by community laws and protocols. This is the digital equivalent of land rights; it is about having a secure, sovereign territory in the digital world where cultural and informational assets can be protected and managed for future generations.

Finally, asserting sovereignty is a political and legal act. It means communities proactively developing and formalizing their own data governance protocols. It means creating clear rules for how they wish to engage with outside researchers, companies, and governments. It means negotiating data sharing agreements that are not boilerplate contracts, but strong, legally binding documents that embed the CARE principles and specify community ownership and control. This is the hard work of turning the principle of sovereignty into a daily, operational reality. It is a declaration that in the 21st century, the struggle for self-determination, cultural preservation, and the well-being of the community will be fought and won not only on the land, but in the digital world as well.

About Author

Azhar ul Haque Sario is a bestselling author and data scientist with a remarkable record of achievement. This Cambridge alumnus brings a wealth of knowledge to his work, holding an MBA, ACCA (Knowledge Level - FTMS College Malaysia and London College of Accountancy, London), BBA, and several Google certifications, including specializations in Google Data Analytics, Google Digital Marketing & E-commerce, and Google Project Management.

With ten years of business experience, Azhar combines practical expertise with his impressive academic background to craft insightful books. His prolific writing has resulted in an astounding 2810 published titles, earning him the record for the maximum Kindle editions and paperback books published by an individual author in one year, awarded by Asia Books of Records in 2024.

ORCID: https://orcid.org/0009-0004-8629-830X
Azhar.sario@hotmail.co.uk
https://www.linkedin.com/in/azharulhaquesario/

www.ingramcontent.com/pod-product-compliance
Lightning Source LLC
Chambersburg PA
CBHW031347230926
51848CB00002B/419